AN INTRODUCTION TO QUEUEING THEORY
AND MATRIX-ANALYTIC METHODS

An Introduction to Queueing Theory and Matrix-Analytic Methods

by

L. BREUER
University of Trier, Germany

and

D. BAUM
University of Trier, Germany

 Springer

A C.I.P. Catalogue record for this book is available from the Library of Congress.

ISBN   978-90-481-6913-9 (PB)
ISBN   978-1-4020-3631-6 (e-book)

Published by Springer,
P.O. Box 17, 3300 AA Dordrecht, The Netherlands.

www.springeronline.com

Printed on acid-free paper

Printed in the Netherlands.

Contents

List of Figures

Foreword

The present textbook contains the records of a two–semester course on queueing theory, including an introduction to matrix–analytic methods. This course comprises four hours of lectures and two hours of exercises per week and has been taught at the University of Trier, Germany, for about ten years in sequence. The course is directed to last year undergraduate and first year graduate students of applied probability and computer science, who have already completed an introduction to probability theory. Its purpose is to present material that is close enough to concrete queueing models and their applications, while providing a sound mathematical foundation for the analysis of these. Thus the goal of the present book is two–fold.

On the one hand, students who are mainly interested in applications easily feel bored by elaborate mathematical questions in the theory of stochastic processes. The presentation of the mathematical foundations in our courses is chosen to cover only the necessary results, which are needed for a solid foundation of the methods of queueing analysis. Further, students oriented towards applications expect to have a justification for their mathematical efforts in terms of immediate use in queueing analysis. This is the main reason why we have decided to introduce new mathematical concepts only when they will be used in the immediate sequel.

On the other hand, students of applied probability do not want any heuristic derivations just for the sake of yielding fast results for the model at hand. They want to see the close connections between queueing theory and the theory of stochastic processes. For them, a systematic introduction to the necessary concepts of Markov renewal theory is indispensable. Further, they are not interested in any technical details of queueing applications, but want to see the reflection of the mathematical concepts in the queueing model as purely as possible.

A prominent part of the book will be devoted to matrix–analytic methods. This is a collection of approaches which extend the applicability of Markov renewal methods to queueing theory by introducing a finite number of auxiliary states. For the embedded Markov chains this leads to transition matrices in block form having the same structure as the classical models. With a few modifications they can be analyzed in the same way.

Matrix–analytic methods have become quite popular in queueing theory during the last twenty years. The intention to include these in a students' introduction to queueing theory has been the main motivation for the authors to write the present book. Its aim is a presentation of the most important matrix–analytic concepts like phase–type distributions, Markovian arrival processes, the GI/PH/1 and BMAP/G/1 queues as well as QBDs and discrete time approaches. This is the content of part III of this book.

As an introductory course for students it is necessary to provide the required results from Markov renewal theory before. This is done in part I, which contains Markovian theory, and part II which combines the concepts of part I with renewal theory in order to obtain a foundation for Markov renewal theory. Certainly only few students would like to acquire this theoretical body without some motivating applications in classical queueing theory. These are introduced as soon as the necessary theoretical background is provided.

The book is organized as follows. The first chapter gives a short overview of the diverse application areas for queueing theory and defines queues and their system processes (number of users in the system). The appendix sections in chapter 15 provide an easy reference to some basic concepts of analysis and probability theory.

For the simple Markovian queueing models (in discrete and continuous time) it suffices to give a short introduction to Markov chains and processes, and then present an analysis of some queueing examples. This is done in chapters 2 through 4. Chapter 5 gives an introduction to the analysis of simple queueing networks, in particular Jackson and Gordon–Newell networks as well as BCMP networks. This concludes the first part of the book, which deals with Markovian methods exclusively.

The second part is devoted to semi–Markovian methods. In chapter 6 the most important results of renewal theory are provided. Chapter 7 contains a short introduction to Markov renewal theory. This will be necessary for the analysis of the classical semi–Markovian queues (namely the GI/M/1 and M/G/1 systems), which is presented in chapter 8.

More recent approaches which are usually subsumed under the term "matrix–analytic methods" are presented in the third part of the book. In chapters

9 and 10 the basic concepts of phase–type distributions and Markovian arrival processes are introduced. The matrix–analytic analogues to the GI/M/1 and M/G/1 queues, namely the GI/PH/1 and BMAP/G/1 systems are analyzed in chapters 11 and 12. Chapter 13 gives a short overview on discrete time analogues. Further blockwise skip–free Markov chains, also known as QBD processes, are analyzed, with an application to the PH/PH/1 queue in discrete time. Finally, in chapter 14 a generalization of BMAPs towards spatial Markovian arrival processes is presented.

Of course, most of the more classical material can be found in existing textbooks on stochastic processes. For example, Çinlar [25] and Ross [75] still contain, in our view, the most systematic treatment of semi–Markovian queues. Also of great value, mostly for the theory of Markov chains and processes, are the courses on stochastic processes by Karlin and Taylor [46, 47]. Further important results may be found in Doob [31], Asmussen [5], and Nelson [61]. The material on queueing networks can be found in Mitrani [60], Kelly [48], and Kleinrock [50]. Monographs on matrix–analytic methods are the pioneering books by Neuts [65, 66], and Latouche and Ramaswami [52]. For discrete time methods the overview paper by Alfa [2] was helpful.

However, some aspects of standard presentation have been changed in order to alleviate the mathematical burden for the students. The stationary regime for Markov chains has been introduced as an asymptotic mean over time in order to avoid the introduction of periodicity of states. The definition of Markov processes in chapter 3 is much closer to the derivation of immediate results. It is not necessary to derive the standard path properties in lengthy preliminary analyses, since these are already included in the definition. Nevertheless, the close connection between the phenomena observed in queueing systems and the definition given in our textbook is immediately clear to the student.

The introduction of renewal theory has been postponed to the second part of the book in order to show a variety of queueing application of a purely Markovian nature first. The drawback that a proof for asymptotic behaviour of Markov processes must be deferred appears bearable for an average student. The proof of Blackwell's theorem, and thus also for the equivalent key renewal theorem, has been omitted as it is too technical for a student presentation in the authors' opinion. The same holds for proofs regarding the necessity of the stability condition for the queues GI/PH/1 and BMAP/G/1. Only proofs for sufficiency have been included because they are easily based on the classical Foster criteria.

At the end of each chapter there will be a collection of exercises, some of them representing necessary auxiliary results to complete the proofs presented in the lectures. Additional material is given as exercises, too, e.g. examples of computer networks or certain special queueing system.

The book is written according to the actual scripts of the lecture courses given at the University of Trier, Germany. It is intended not only to collect material which can be used for an introductory course on queueing theory, but to propose the scripts of the lectures themselves. The book contains exactly as much material as the authors (as lecturers) could present in two semesters. Thus a lecturer using this textbook does not need to choose and reassemble the material for a course from sources which must be shortened because there is no time to treat them completely. This entails saving the work of reformulating notations and checking dependencies. For a course of only one semester we propose to teach parts I and II of this book, leaving out sections 5.3 and 8.3.

Chapter 1

QUEUES: THE ART OF MODELLING

Stochastic modelling is the application of probability theory to the description and analysis of real world phenomena. It is thus comparable to physics, with the distinguishing property that mostly technical and not natural systems are investigated. These are usually so complex that deterministic laws cannot be formulated, a circumstance that leads to pervasive use of stochastic concepts. Application fields as telecommunication or insurance bring methods and results of stochastic modelling to the attention of applied sciences such as engineering or economics. On the other hand, often new technological developments give rise to new questions in an application field, which in turn may open a new direction in stochastic research, and thus provide an impetus to applied probability. Stochastic modelling is a science with close interaction between theory and practical applications. This is nice because it combines the possibility of theoretical beauty with a real–world meaning of its key concepts. On the other hand, it is difficult to cover the whole width from theoretical foundations to the details of practical applications. The present book is an essay to give an introduction to the theory of stochastic modelling in a systematic way without losing contact to its applicability.

One of the most important domains in stochastic modelling is the field of queueing theory. This shall be the topic of this treatise. Many real systems can be reduced to components which can be modelled by the concept of a so–called queue. The basic idea of this concept has been borrowed from the every–day experience of the queues at the checkout counters in a supermarket. A queue in the more exact scientific sense consists of a system into which there comes a stream of users who demand some capacity of the system over a certain time interval before they leave the system again. It is said that the users are served in the system by one or many servers. Thus a queueing system can be described by a (stochastic) specification of the arrival stream and of the system demand for every user as well as a definition of the service mechanism. The former describe the input into a queue, while the latter represents the functioning of the inner mechanisms of a queueing system. Before we give an exact definition of a queueing system, a few examples shall provide an idea of the variety of applications.

Example 1.1 Single Server Queue

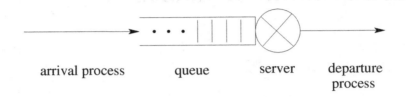

arrival process   queue   server   departure
                                                        process

Figure 1.1.   Single server queue

A queue in front of the checkout counter of a supermarket may serve as the simplest illustration for a queueing system. There is one input stream, and one server who serves the customers in order of their appearance at the counter. This service discipline, which does not admit any preferences among users, is called first come first served (abbr.: FCFS).

Example 1.2 Multi–Server Queue

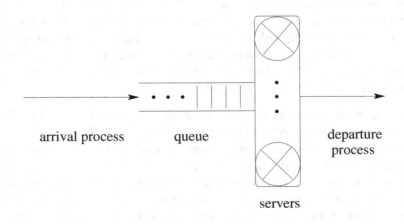

arrival process   queue   departure
                                                        process

servers

Figure 1.2.   Multi–server queue

The first real application of queueing theory, in fact the one that engendered the development of the whole field of research, has been the design and analysis of telephone networks. In the early days of Erlang at the beginning of the 20th century, telephone calls first went to an operator before they could be conected to the person that was to be reached by the call. Thus an important part of a telephone network could be modelled by a queueing system in which the servers are the operators in a call center who connect the incoming calls (which are modelled by the input stream of users) to their addressees. Here, the time of connecting is represented by the service demand of a user. A crucial performance measure of such a system is the probability that a person who wants to get a connection for a call finds all operators busy and thus cannot be served. This value is called the loss probability of the system. For a modern call center, where questions are answered instead of cables connected, the service times represent the time of the call between the user and the operator.

Example 1.3 In recent times, computer networks (the most prominent example is the internet) have increasingly become the object of applications of queueing theory. For example, a server in a computer network receives demands from its clients and needs to serve them. The demands make up the input stream into the queueing sytem that represents the server utilization. A service discipline that is often used in these kinds of application is the following: The processing capacity of the server is divided into equal parts among the jobs such that none of the clients is favoured, but each client's service time depends on the total number of clients that are resident at the same time. Because of its prevalence in computer applications, this service discipline is called processor sharing.

Example 1.4 Queues find further applications in airport traffic. Here, the servers are the several landing fields available for arriving airplanes, while the latter are the users of the system. Obviously, there cannot be any queue of planes waiting in the air, so that an arriving airplane finding all landing fields in use needs instead to fly an extra circle around the airport and then try again for a possibility to land. Such a manoeuver is called a retrial, and the corresponding queueing model is called a retrial queue. Since with every extra circle that a plane has to perform its gasoline is reduced more, the priority of such an aircraft to obtain a landing permission is increasing and should be higher than that of more recent airplanes with fewer retrials. Such an influence on the service schedule is called priority queueing.

Example 1.5 More complicated queueing models have been developed for the design of traffic lights at crossroads. In such a model, there are several distinguishable queues which represent the different roads leading to the intersection. A green light at one road means that vehicles waiting on it are served on a first come first served base. There are as many servers as there are traffic lights at the intersection, and it is obvious that these servers must function in dependence on each other. Such queueing systems are called polling systems.

Example 1.6 In modern production systems an analysis of assembly lines has become a fundamental necessity. They are modelled by so–called tandem queueing networks, which are defined as a series of several single queueing systems where the output of one queue forms the input of the next.

Example 1.7 Finally, perhaps the most interesting object of analysis for to-day's computer science, the internet, would merely appear as a highly complex queueing network, at least so from the point of view of stochastic modelling.

These examples illustrate the very different interpretations and thus applications that queueing systems can assume. They should suffice as a motivation to undergo the strain of developing methods and concepts for the analysis of queueing systems of the highest possible complexity and generality. Our introduction to the theory of queues gives a (hopefully) balanced presentation of potentially very general methods of analysis based on the theory of Markov renewal processes, and at the same time tries to apply these to the practically relevant analyses of queueing systems. Opening the exact part of the presentation we begin with a definition of the concept of a queue:

For every $n \in \mathbb{N}$, let T_n and S_n denote positive real–valued random variables with $T_{n+1} > T_n$ for all $n \in \mathbb{N}$. The sequence $\mathcal{T} = (T_n : n \in \mathbb{N}_0)$ is called **arrival point process** and $\mathcal{S} = (S_n : n \in \mathbb{N})$ is the sequence of **service times**. Further choose a number k of servers and the **system capacity** c, with $k, c \in \mathbb{N} \cup \{\infty\}$.

Finally a **service discipline** B needs to be specified. This can be first come first served (FCFS), last come first served (LCFS), processor sharing (PS), sometimes working with certain priorities or preemption rules. Normally we choose FCFS, meaning that the first user who arrives in the system will be the first to get access to a server. If other service disciplines will be used, they will be explained whenever introduced.

The 5–tuple $(\mathcal{T}, \mathcal{S}, k, c, B)$ is called a **queue** (or **queueing system**) with arrival point process $\mathcal{T}$, sequence of service times $\mathcal{S}$, number k of servers, system capacity c, and service discipline B.

Define further the nth **inter–arrival time** by $Z_1 := T_1$ and $Z_n := T_n - T_{n-1}$ for all $n \geq 2$. The standard way to specify a queue is the **Kendall notation**. This merely denotes the 5–tuple $(\mathcal{T}, \mathcal{S}, k, c, B)$ in the above definition by $\mathcal{T}/\mathcal{S}/k/c/B$ and additionally sets some conventions for interpreting this notation: If the 4th or 5th parameter is left out, this is agreed to mean $c = \infty$ or $B = FCFS$, respectively. Further, for the first two parameters the letters M (resp. G) stand for geometric (resp. general) inter–arrival and service times for queues in discrete time and for exponential (resp. general) inter–arrival and service times for queues in continuous time. There are additional conventions: D stands for deterministic (Dirac) distributions, Geo is the same as M for discrete time systems, etc.

The main goal of any queueing analysis will be to specify and analyze the **system process** $\mathcal{Q} = (Q_t : t \geq 0)$, where Q_t is the number of users in the queueing system (usually including the number of users in service) at time t. An important measure (in case of existence) will be the **asymptotic distribution** of Q_t for t tending to infinity.

Our first result concerns a sample path property of general conservative systems with inputs and outputs. Conservative systems do not create or destroy users. Let $\alpha(t)$ denote the number of arrivals into the system until time t. Define $\lambda_t := \alpha(t)/t$ as the average arrival rate during the interval $[0, t]$. Further define T as the average time a user spends in the system. Finally denote the average number of users in the system during $[0, t]$ by $\bar{N}_t$. Then we can state

Theorem 1.8 *Little's Result*
If the limit $\lambda = \lim_{t \to \infty} \lambda_t$ and T do exist, then the limit $\bar{N} = \lim_{t \to \infty} \bar{N}_t$ does exist, too, and the relation
$$\bar{N} = \lambda T$$
holds.

Proof: We introduce the notation $\delta(t)$ for the number of departures from the system during $[0, t]$ and $N(t)$ for the number of users in the system. If the system starts empty, then these definitions imply the relation

$$N(t) = \alpha(t) - \delta(t)$$

for all times t (see the following figure).

Denote the total time that all users have spent in the system during $[0, t]$ by

$$\gamma(t) := \int_0^t N(s) \, ds$$

If we define T_t as the system time per user averaged over all users in the interval $[0, t]$, then the definitions of $\alpha(t)$ and $\gamma(t)$ imply the relation

$$T_t = \frac{\gamma(t)}{\alpha(t)} \tag{1.1}$$

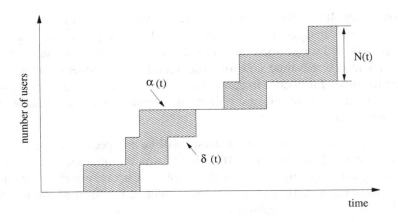

Figure 1.3.   Total system time

The average number of users in the system during $[0, t]$ can be obtained as

$$\bar{N}_t = \frac{\gamma(t)}{t} = \frac{\gamma(t)}{\alpha(t)} \cdot \frac{\alpha(t)}{t} = \lambda_t T_t$$

where the last equality comes from (1.1) and the definition of λ_t. If the limits λ and $T = \lim_{t \to \infty} T_t$ exist, then the stated relation $\bar{N} = \lambda T$ follows for t tending to infinity.
□

For ease of reference, we finally provide a table of some basic probability distributions which will occur frequently throughout the book.

Distribution	Density	Range	Parameters
Exponential	$\lambda e^{-\lambda t}$	$t > 0$	$\lambda > 0$
Erlang	$\frac{m\mu(m\mu t)^{m-1}}{(m-1)!} e^{-m\mu t}$	$t > 0$	$m \in \mathbb{N}, \mu > 0$
Poisson	$\frac{\lambda^n}{n!} e^{-\lambda}$	$n \in \mathbb{N}_0$	$\lambda > 0$
Geometric	$(1-p)p^n$	$n \in \mathbb{N}_0$	$p \in]0, 1[$
Binomial	$\binom{N}{n} p^n (1-p)^{N-n}$	$0 \leq n \leq N$	$N \in \mathbb{N}, p \in]0, 1[$

Notes

The first formal proof for Little's result appeared in Little [53]. The proof presented here is taken from Kleinrock [50].

PART I

MARKOVIAN METHODS

Chapter 2

MARKOV CHAINS AND QUEUES IN DISCRETE TIME

1. Definition

Let X_n with $n \in \mathbb{N}_0$ denote random variables on a discrete space E. The sequence $\mathcal{X} = (X_n : n \in \mathbb{N}_0)$ is called a **stochastic chain**. If $\mathbb{P}$ is a probability measure $\mathcal{X}$ such that

$$\mathbb{P}(X_{n+1} = j | X_0 = i_0, \dots, X_n = i_n) = \mathbb{P}(X_{n+1} = j | X_n = i_n) \quad (2.1)$$

for all $i_0, \dots, i_n, j \in E$ and $n \in \mathbb{N}_0$, then the sequence $\mathcal{X}$ shall be called a **Markov chain** on E. The probability measure $\mathbb{P}$ is called the distribution of $\mathcal{X}$, and E is called the **state space** of $\mathcal{X}$.

If the conditional probabilities $\mathbb{P}(X_{n+1} = j | X_n = i_n)$ are independent of the time index $n \in \mathbb{N}_0$, then we call the Markov chain $\mathcal{X}$ **homogeneous** and denote

$$p_{ij} := \mathbb{P}(X_{n+1} = j | X_n = i)$$

for all $i, j \in E$. The probability p_{ij} is called **transition probability** from state i to state j. The matrix $P := (p_{ij})_{i,j \in E}$ shall be called **transition matrix** of the chain $\mathcal{X}$. Condition (2.1) is referred to as the **Markov property**.

Example 2.1 If $(X_n : n \in \mathbb{N}_0)$ are random variables on a discrete space E, which are stochastically independent and identically distributed (shortly: iid), then the chain $\mathcal{X} = (X_n : n \in \mathbb{N}_0)$ is a homogeneous Markov chain.

Example 2.2 Discrete Random Walk
Set $E := \mathbb{Z}$ and let $(S_n : n \in \mathbb{N})$ be a sequence of iid random variables with values in $\mathbb{Z}$ and distribution π. Define $X_0 := 0$ and $X_n := \sum_{k=1}^{n} S_k$ for all

$n \in \mathbb{N}$. Then the chain $\mathcal{X} = (X_n : n \in \mathbb{N}_0)$ is a homogeneous Markov chain with transition probabilities $p_{ij} = \pi_{j-i}$. This chain is called **discrete random walk**.

Example 2.3 Bernoulli process
Set $E := \mathbb{N}_0$ and choose any parameter $0 < p < 1$. The definitions $X_0 := 0$ as well as

$$p_{ij} := \begin{cases} p, & j = i + 1 \\ 1 - p, & j = i \end{cases}$$

for $i \in \mathbb{N}_0$ determine a homogeneous Markov chain $\mathcal{X} = (X_n : n \in \mathbb{N}_0)$. It is called **Bernoulli process** with parameter p.

So far, al examples have been chosen as to be homogeneous. The following theorem shows that there is a good reason for this:

Theorem 2.4 *Be* $\mathcal{X} = (X_n : n \in \mathbb{N}_0)$ *a Markov chain on a discrete state space* E. *Then there is a homogeneous Markov chain* $\mathcal{X}' = (X_n' : n \in \mathbb{N}_0)$ *on the state space* $E \times \mathbb{N}_0$ *such that* $X_n = pr_1(X_n')$ *for all* $n \in \mathbb{N}_0$, *with* pr_1 *denoting the projection to the first dimension.*

Proof: Let $\mathcal{X}$ be a Markov chain with transition probabilities

$$p_{n;ij} := \mathbb{P}(X_{n+1} = j | X_n = i)$$

which may depend on the time instant n. Define the two–dimensional random variables $X_n' := (X_n, n)$ for all $n \in \mathbb{N}_0$ and denote the resulting distribution of the chain $\mathcal{X}' = (X_n' : n \in \mathbb{N}_0)$ by $\mathbb{P}'$. By definition we obtain $X_n = pr_1(X_n')$ for all $n \in \mathbb{N}_0$.

Further $\mathbb{P}'(X_0' = (i,k)) = \delta_{k0} \cdot \mathbb{P}(X_0 = i)$ holds for all $i \in E$, and all transition probabilities

$$p_{(i,k),(j,l)}' = \mathbb{P}'(X_{k+1}' = (j,l) | X_k' = (i,k)) = \delta_{l,k+1} \cdot p_{k;ij}$$

can be expressed without a time index. Hence the Markov chain $\mathcal{X}'$ is homogeneous.
$\square$

Because of this result, we will from now on treat only homogeneous Markov chains and omit the adjective "homogeneous".

Let P denote the transition matrix of a Markov chain on E. Then as an immediate consequence of its definition we obtain $p_{ij} \in [0, 1]$ for all $i, j \in E$

and $\sum_{j \in E} p_{ij} = 1$ for all $i \in E$. A matrix P with these properties is called a **stochastic matrix** on E. In the following we shall demonstrate that, given an initial distribution, a Markov chain is uniquely determined by its transition matrix. Thus any stochastic matrix defines a family of Markov chains.

Theorem 2.5 *Let $\mathcal{X}$ denote a homogeneous Markov chain on E with transition matrix P. Then the relation*

$$\mathbb{P}(X_{n+1} = j_1, \dots, X_{n+m} = j_m | X_n = i) = p_{i,j_1} \cdot \dots \cdot p_{j_{m-1},j_m}$$

holds for all $n \in \mathbb{N}_0$, $m \in \mathbb{N}$, and $i, j_1, \dots, j_m \in E$.

Proof: This is easily shown by induction on m. For $m = 1$ the statement holds by definition of P. For $m > 1$ we can write

$$\mathbb{P}(X_{n+1} = j_1, \dots, X_{n+m} = j_m | X_n = i)$$
$$= \frac{\mathbb{P}(X_{n+1} = j_1, \dots, X_{n+m} = j_m, X_n = i)}{\mathbb{P}(X_n = i)}$$
$$= \frac{\mathbb{P}(X_{n+1} = j_1, \dots, X_{n+m} = j_m, X_n = i)}{\mathbb{P}(X_{n+1} = j_1, \dots, X_{n+m-1} = j_{m-1}, X_n = i)}$$
$$\times \frac{\mathbb{P}(X_{n+1} = j_1, \dots, X_{n+m-1} = j_{m-1}, X_n = i)}{\mathbb{P}(X_n = i)}$$
$$= \mathbb{P}(X_{n+m} = j_m | X_n = i, X_{n+1} = j_1, \dots, X_{n+m-1} = j_{m-1})$$
$$\times p_{i,j_1} \cdot \dots \cdot p_{j_{m-2},j_{m-1}}$$
$$= p_{j_{m-1},j_m} \cdot p_{i,j_1} \cdot \dots \cdot p_{j_{m-2},j_{m-1}}$$

because of the induction hypothesis and the Markov property.
□

Let π be a probability distribution on E with $\mathbb{P}(X_0 = i) = \pi_i$ for all $i \in E$. Then theorem 2.5 immediately yields

$$\mathbb{P}(X_0 = j_0, X_1 = j_1, \dots, X_m = j_m) = \pi_{j_0} \cdot p_{j_0,j_1} \cdots p_{j_{m-1},j_m} \qquad (2.2)$$

for all $m \in \mathbb{N}$ and $j_0, \dots, j_m \in E$. The chain with this distribution $\mathbb{P}$ is denoted by $\mathcal{X}^\pi$ and called the π–**version** of $\mathcal{X}$. The probability measure π is called **initial distribution** for $\mathcal{X}$.

Theorem 2.5 and the extension theorem by Tulcea (see appendix 2) show that a Markov chain is uniquely determined by its transition matrix and its initial distribution. Whenever the initial distribution π is not important or understood from the context, we will simply write $\mathcal{X}$ instead of $\mathcal{X}^\pi$. However, in an exact manner the notation $\mathcal{X}$ denotes the family of all the versions $\mathcal{X}^\pi$ of $\mathcal{X}$, indexed by their initial distribution π.

Theorem 2.6 *Let $\mathcal{X}$ denote a homogeneous Markov chain with transition matrix P. Then the relation*

$$\mathbb{P}(X_{n+m} = j | X_n = i) = P^m(i, j)$$

holds for all $m, n \in \mathbb{N}_0$ and $i, j \in E$, with $P^m(i, j)$ denoting the (i, j)th entry of the mth power of the matrix P. In particular, P^0 equals the identity matrix.

Proof: This follows by induction on m. For $m = 1$ the statement holds by definition of P. For $m > 1$ we can write

$$
\begin{aligned}
\mathbb{P}(X_{n+m} = j | X_n = i) &= \frac{\mathbb{P}\left(X_{n+m} = j, X_n = i\right)}{\mathbb{P}\left(X_n = i\right)} \\
&= \sum_{k \in E} \frac{\mathbb{P}\left(X_{n+m} = j, X_{n+m-1} = k, X_n = i\right)}{\mathbb{P}\left(X_{n+m-1} = k, X_n = i\right)} \\
&\qquad \times \frac{\mathbb{P}\left(X_{n+m-1} = k, X_n = i\right)}{\mathbb{P}\left(X_n = i\right)} \\
&= \sum_{k \in E} \mathbb{P}\left(X_{n+m} = j | X_{n+m-1} = k, X_n = i\right) \cdot P^{m-1}(i, k) \\
&= \sum_{k \in E} p_{kj} \cdot P^{m-1}(i, k) = P^m(i, j)
\end{aligned}
$$

because of the induction hypothesis and the Markov property.
□

Thus the probabilities for transitions in m steps are given by the mth power of the transition matrix P. The rule $P^{m+n} = P^m P^n$ for the multiplication of matrices and theorem 2.6 lead to the decompositions

$$\mathbb{P}(X_{m+n} = j | X_0 = i) = \sum_{k \in E} \mathbb{P}(X_m = k | X_0 = i) \cdot \mathbb{P}(X_n = j | X_0 = k)$$

which are known as the **Chapman–Kolmogorov equations**.

For later purposes we will need a relation closely related to the Markov property, which is called the **strong Markov property**. Let τ denote a random variable with values in $\mathbb{N}_0 \cup \{\infty\}$, such that the condition

$$\mathbb{P}(\tau \le n | \mathcal{X}) = \mathbb{P}(\tau \le n | X_0, \dots, X_n) \qquad (2.3)$$

holds for all $n \in \mathbb{N}_0$. Such a random variable is called a (discrete) **stopping time** for $\mathcal{X}$. The defining condition means that the probability for the event $\{\tau \le n\}$ depends only on the evolution of the chain until time n. In other words, the determination of a stopping time does not require any knowledge of the future. Now the strong Markov property is stated in

Theorem 2.7 *Let $\mathcal{X}$ denote a Markov chain and τ a stopping time for $\mathcal{X}$ with $\mathbb{P}(\tau < \infty) = 1$. Then the relation*

$$\mathbb{P}(X_{\tau+m} = j | X_0 = i_0, \ldots, X_\tau = i_\tau) = \mathbb{P}(X_m = j | X_0 = i_\tau)$$

holds for all $m \in \mathbb{N}$ and $i_0, \ldots, i_\tau, j \in E$.

Proof: The fact that the stopping time τ is finite and may assume only countably many values can be exploited in the transformation

$$\mathbb{P}(X_{\tau+m} = j | X_0 = i_0, \ldots, X_\tau = i_\tau)$$

$$= \sum_{n=0}^{\infty} \mathbb{P}(\tau = n, X_{\tau+m} = j | X_0 = i_0, \ldots, X_\tau = i_\tau)$$

$$= \sum_{n=0}^{\infty} \mathbb{P}(X_{\tau+m} = j | \tau = n, X_0 = i_0, \ldots, X_\tau = i_\tau)$$

$$\times \mathbb{P}(\tau = n | X_0 = i_0, \ldots, X_\tau = i_\tau)$$

$$= \sum_{n=0}^{\infty} \mathbb{P}(X_{n+m} = j | X_n = i_\tau) \cdot \mathbb{P}(\tau = n | \mathcal{X})$$

$$= \sum_{n=0}^{\infty} \mathbb{P}(\tau = n | \mathcal{X}) \cdot \mathbb{P}(X_m = j | X_0 = i_\tau)$$

which yields the statement, as τ is finite with probability one.
□

2. Classification of States

Let $\mathcal{X}$ denote a Markov chain with state space E and transition matrix P. We call a state $j \in E$ **accessible** from a state $i \in E$ if there is a number $m \in \mathbb{N}_0$ with $P(X_m = j | X_0 = i) > 0$. This relation shall be denoted by $i \rightarrow j$. If for two states $i, j \in E$, the relations $i \rightarrow j$ and $j \rightarrow i$ hold, then i and j are said to **communicate**, in notation $i \leftrightarrow j$.

Theorem 2.8 *The relation $\leftrightarrow$ of communication between states is an equivalence relation.*

Proof: Because of $P^0 = I$, communication is reflexive. Symmetry holds by definition. Thus it remains to show transitivity. For this, assume $i \leftrightarrow j$

and $j \leftrightarrow k$ for three states $i, j, k \in E$. This means that there are numbers $m, n \in \mathbb{N}_0$ with $P^m(i,j) > 0$ and $P^n(j,k) > 0$. Hence, by the Chapman–Kolmogorov equation, we obtain

$$\mathbb{P}(X_{m+n} = k | X_0 = i) = \sum_{h \in E} \mathbb{P}(X_m = h | X_0 = i) \cdot \mathbb{P}(X_n = k | X_0 = h)$$

$$\geq \mathbb{P}(X_m = j | X_0 = i) \cdot \mathbb{P}(X_n = k | X_0 = j) > 0$$

which proves $i \rightarrow k$. The remaining proof of $k \rightarrow i$ is completely analogous.
□

Because of this result and the countability, we can divide the state space E of a Markov chain into a partition of countably many equivalence classes with respect to the communication of states. Any such equivalence class shall be called **communication class**. A communication class $C \subset E$ that does not allow access to states outside itself, i.e. for which the implication

$$i \rightarrow j, \quad i \in C \quad \Rightarrow \quad j \in C$$

holds, is called **closed**. If a closed equivalence class consists only of one state, then this state shall be called **absorbing**. If a Markov chain has only one communication class, i.e. if all states are communicating, then it is called **irreducible**. Otherwise it is called **reducible**.

Example 2.9 Let $\mathcal{X}$ denote a discrete random walk (see example 2.2) with the specification $\pi_1 = p$ and $\pi_{-1} = 1 - p$ for some parameter $0 < p < 1$. Then $\mathcal{X}$ is irreducible.

Example 2.10 The Bernoulli process (see example 2.3) with non–trivial parameter $0 < p < 1$ is to the highest degree reducible. Every state $x \in \mathbb{N}_0$ forms an own communication class. None of these is closed, thus there are no absorbing states.

Theorem 2.11 *Be $\mathcal{X}$ a Markov chain with state space E and transition matrix P. Let $C = \{c_n : n \in I\} \subset E$ with $I \subset \mathbb{N}$ be a closed communication class. Define the matrix P' by its entries $p'_{ij} := p_{c_i, c_j}$ for all $i, j \in I$. Then P' is stochastic.*

Proof: By definition, $p'_{ij} \in [0, 1]$ for all $i, j \in I$. Since C is closed, $p_{c_i, k} = 0$ for all $i \in I$ and $k \notin C$. This implies

$$\sum_{j \in I} p'_{ij} = \sum_{j \in I} p_{c_i, c_j} = 1 - \sum_{k \notin C} p_{c_i, k} = 1$$

for all $i \in I$, as P is stochastic.
□

Thus the restriction of a Markov chain $\mathcal{X}$ with state space E to the states of one of its closed communication classes C defines a new Markov chain with state space C. If the states are relabeled according to their affiliation to a communication class, the transition matrix of $\mathcal{X}$ can be displayed in a block matrix form as

$$
P = \begin{bmatrix}
Q & Q_1 & Q_2 & Q_3 & Q_4 & \cdots \\
0 & P_1 & 0 & 0 & 0 & \cdots \\
0 & 0 & P_2 & 0 & 0 & \cdots \\
0 & 0 & 0 & P_3 & 0 & \cdots \\
\vdots & \vdots & \ddots & \ddots & \ddots &
\end{bmatrix}
\tag{2.4}
$$

with P_n being stochastic matrices on the closed communication classes C_n. The first row contains the transition probabilities starting from communication classes that are not closed.

Let $\mathcal{X}$ denote a Markov chain with state space E. In the rest of this section we shall investigate distribution and expectation of the following random variables: Define τ_j as the stopping time of the **first visit** to the state $j \in E$, i.e.

$$
\tau_j := \min\{n \in \mathbb{N} : X_n = j\}
$$

Denote the distribution of τ_j by

$$
F_k(i,j) := \mathbb{P}(\tau_j = k | X_0 = i)
$$

for all $i, j \in E$ and $k \in \mathbb{N}$.

Lemma 2.12 *The conditional distribution of the first visit to the state $j \in E$, given an initial state $X_0 = i$, can be determined iteratively by*

$$
F_k(i,j) = \begin{cases}
p_{ij}, & k = 1 \\
\sum_{h \neq j} p_{ih} F_{k-1}(h,j), & k \geq 2
\end{cases}
$$

for all $i, j \in E$.

Proof: For $k = 1$, the definition yields

$$
F_1(i,j) = \mathbb{P}(\tau_j = 1 | X_0 = i) = \mathbb{P}(X_1 = j | X_0 = i) = p_{ij}
$$

for all $i, j \in E$. For $k \geq 2$, conditioning upon X_1 yields

$$
F_k(i,j) = \mathbb{P}(X_1 \neq j, \ldots, X_{k-1} \neq j, X_k = j | X_0 = i)
$$

$$= \sum_{h \neq j} \mathbb{P}(X_1 = h | X_0 = i)$$

$$\times \mathbb{P}(X_2 \neq j, \dots, X_{k-1} \neq j, X_k = j | X_0 = i, X_1 = h)$$

$$= \sum_{h \neq j} p_{ih} \cdot \mathbb{P}(X_1 \neq j, \dots, X_{k-2} \neq j, X_{k-1} = j | X_0 = h)$$

due to the Markov property.

$\square$

Now define

$$f_{ij} := \mathbb{P}(\tau_j < \infty | X_0 = i) = \sum_{k=1}^{\infty} F_k(i, j) \qquad (2.5)$$

for all $i, j \in E$, which represents the probability of ever visiting state j after beginning in state i. Summing up over all $k \in \mathbb{N}$ in the formula of Lemma 2.12 leads to

$$f_{ij} = p_{ij} + \sum_{h \neq j} p_{ih} f_{hj} \qquad (2.6)$$

for all $i, j \in E$. The proof is left as an exercise.

Define N_j as the random variable of the **total number of visits** to the state $j \in E$. Expression (2.6) is useful for computing the distribution of N_j:

Theorem 2.13 *Let $\mathcal{X}$ denote a Markov chain with state space E. The total number of visits to a state $j \in E$ under the condition that the chain starts in state i is given by*

$$\mathbb{P}(N_j = m | X_0 = j) = f_{jj}^{m-1}(1 - f_{jj})$$

and for $i \neq j$

$$\mathbb{P}(N_j = m | X_0 = i) = \begin{cases} 1 - f_{ij}, & m = 0 \\ f_{ij} f_{jj}^{m-1}(1 - f_{jj}), & m \geq 1 \end{cases}$$

Thus the distribution of N_j is modified geometric.

Proof: Define $\tau_j^{(1)} := \tau_j$ and $\tau_j^{(k+1)} := \min\{n > \tau_j^{(k)} : X_n = j\}$ for all $k \in \mathbb{N}$, with the convention that $\min \emptyset = \infty$. Note that $\tau_j^{(k)} = \infty$ implies $\tau_j^{(l)} = \infty$ for all $l > k$.

Then the sequence $(\tau_j^{(k)} : k \in \mathbb{N})$ is a sequence of stopping times. The event $\{N_j = m\}$ is the same as the intersection of the events $\{\tau_j^{(k)} < \infty\}$ for

$k = 1, \ldots, M$ and $\{\tau_j^{(M+1)} = \infty\}$, with $M = m$ if $i \neq j$ and $M = m - 1$ if $i = j$. Now this event can be further described by the intersection of the events $\{\tau_j^{(k+1)} - \tau_j^{(k)} < \infty\}$ for $k = 0, \ldots, M - 1$ and $\{\tau_j^{(M+1)} - \tau_j^{(M)} = \infty\}$, with M as above and the convention $\tau_j^{(0)} := 0$.

The subevent $\{\tau_j^{(k+1)} - \tau_j^{(k)} < \infty\}$ has probability f_{ij} for $k = 0$ and because of the strong Markov property (see theorem 2.7) probability f_{jj} for $k > 0$. The probability for $\{\tau_j^{(M+1)} - \tau_j^{(M)} = \infty\}$ is $1 - f_{ij}$ for $M = 0$ and $1 - f_{jj}$ for $M > 0$. Once more the strong Markov property is the reason for independence of the subevents. Now multiplication of the probabilities leads to the formulae in the statement.
$\square$

Summing over all m in the above theorem leads to

Corollary 2.14 *For all $j \in E$, the zero–one law*

$$\mathbb{P}(N_j < \infty | X_0 = j) = \begin{cases} 1, & f_{jj} < 1 \\ 0, & f_{jj} = 1 \end{cases}$$

holds, i.e. depending on f_{jj} there are almost certainly infinitely many visits to a state $j \in E$.

This result gives rise to the following definitions: A state $j \in E$ is called **recurrent** if $f_{jj} = 1$ and **transient** otherwise. Let us further define the **potential matrix** $R = (r_{ij})_{i,j \in E}$ of the Markov chain by its entries

$$r_{ij} := \mathbb{E}(N_j | X_0 = i)$$

for all $i, j \in E$. Thus an entry r_{ij} gives the expected number of visits to the state $j \in E$ under the condition that the chain starts in state $i \in E$. As such, r_{ij} can be computed by

$$r_{ij} = \sum_{n=0}^{\infty} P^n(i, j) \tag{2.7}$$

for all $i, j \in E$. The results in theorem 2.13 and corollary 2.14 yield

Corollary 2.15 *For all $i, j \in E$ the relations*

$$r_{jj} = (1 - f_{jj})^{-1} \quad \text{and} \quad r_{ij} = f_{ij} r_{jj}$$

hold, with the conventions $0^{-1} := \infty$ and $0 \cdot \infty := 0$ included. In particular, the expected number r_{jj} of visits to the state $j \in E$ is finite if j is transient and infinite if j is recurrent.

Theorem 2.16 *Recurrence and transience of states are class properties with respect to the relation* ↔. *Furthermore, a recurrent communication class is always closed.*

Proof: Assume that $i \in E$ is transient and $i \leftrightarrow j$. Then there are numbers $m, n \in \mathbb{N}$ with $0 < P^m(i, j) \leq 1$ and $0 < P^n(j, i) \leq 1$. The inequalities

$$\sum_{k=0}^{\infty} P^k(i, i) \geq \sum_{h=0}^{\infty} P^{m+h+n}(i, i) \geq P^m(i, j) P^n(j, i) \sum_{k=0}^{\infty} P^k(j, j)$$

now imply $r_{jj} < \infty$ because of representation (2.7). According to corollary 2.15 this means that j is transient, too.

If j is recurrent, then the same inequalities lead to

$$r_{ii} \geq P^m(i, j) P^n(j, i) r_{jj} = \infty$$

which signifies that i is recurrent, too. Since the above arguments are symmetric in i and j, the proof of the first statement is complete.

For the second statement assume that $i \in E$ belongs to a communication class $C \subset E$ and $p_{ij} > 0$ for some state $j \in E \setminus C$. Then

$$f_{ii} = p_{ii} + \sum_{h \neq i} p_{ih} f_{hi} \leq 1 - p_{ij} < 1$$

according to formula (2.6), since $f_{ji} = 0$ (otherwise $i \leftrightarrow j$). Thus i is transient, which proves the second statement.
□

Theorem 2.17 *If the state* $j \in E$ *is transient, then* $\lim_{n \to \infty} P^n(i, j) = 0$, *regardless of the initial state* $i \in E$.

Proof: If the state j is transient, then the first equation in corollary 2.15 yields $r_{jj} < \infty$. The second equation in the same corollary now implies $r_{ij} < \infty$, which by the representation (2.7) completes the proof.
□

3. Stationary Distributions

Let $\mathcal{X}$ denote a Markov chain with state space E and π a measure on E. If $\mathbb{P}(X_n = i) = \mathbb{P}(X_0 = i) = \pi_i$ for all $n \in \mathbb{N}$ and $i \in E$, then $\mathcal{X}^\pi$ is called **stationary**, and π is called a **stationary measure** for $\mathcal{X}$. If furthermore π is a probability measure, then it is called **stationary distribution** for $\mathcal{X}$.

Theorem 2.18 *Let $\mathcal{X}$ denote a Markov chain with state space E and transition matrix P. Further, let π denote a probability distribution on E with $\pi P = \pi$, i.e.*

$$\pi_i = \sum_{j \in E} \pi_j p_{ji} \quad \text{and} \quad \sum_{j \in E} \pi_j = 1$$

for all $i \in E$. Then π is a stationary distribution for X. If π is a stationary distribution for $\mathcal{X}$, then $\pi P = \pi$ holds.

Proof: Let $\mathbb{P}(X_0 = i) = \pi_i$ for all $i \in E$. Then $\mathbb{P}(X_n = i) = \mathbb{P}(X_0 = i)$ for all $n \in \mathbb{N}$ and $i \in E$ follows by induction on n. The case $n = 1$ holds by assumption, and the induction step follows by induction hypothesis and the Markov property. The last statement is obvious.
□

The following examples show some features of stationary distributions:

Example 2.19 Let the transition matrix of a Markov chain $\mathcal{X}$ be given by

$$P = \begin{pmatrix} 0.8 & 0.2 & 0 & 0 \\ 0.2 & 0.8 & 0 & 0 \\ 0 & 0 & 0.4 & 0.6 \\ 0 & 0 & 0.6 & 0.4 \end{pmatrix}$$

Then $\pi = (0.5, 0.5, 0, 0)$, $\pi' = (0, 0, 0.5, 0.5)$ as well as any linear combination of them are stationary distributions for $\mathcal{X}$. This shows that a stationary distribution does not need to be unique.

Example 2.20 Bernoulli process (see example 2.1)
The transition matrix of a Bernoulli process has the structure

$$P = \begin{pmatrix} 1-p & p & 0 & 0 & \cdots \\ 0 & 1-p & p & 0 & \ddots \\ 0 & 0 & 1-p & p & \ddots \\ \vdots & \ddots & \ddots & \ddots & \ddots \end{pmatrix}$$

Hence $\pi P = \pi$ implies first

$$\pi_0 \cdot (1-p) = \pi_0 \quad \Rightarrow \quad \pi_0 = 0$$

since $0 < p < 1$. Assume that $\pi_n = 0$ for any $n \in \mathbb{N}_0$. This and the condition $\pi P = \pi$ further imply for π_{n+1}

$$\pi_n \cdot p + \pi_{n+1} \cdot (1-p) = \pi_{n+1} \quad \Rightarrow \quad \pi_{n+1} = 0$$

which completes an induction argument proving $\pi_n = 0$ for all $n \in \mathbb{N}_0$. Hence the Bernoulli process does not have a stationary distribution.

Example 2.21 The solution of $\pi P = \pi$ and $\sum_{j \in E} \pi_j = 1$ is unique for

$$P = \begin{pmatrix} 1 - p & p \\ p & 1 - p \end{pmatrix}$$

with $0 < p < 1$. Thus there are transition matrices which have exactly one stationary distribution.

The question of existence and uniqueness of a stationary distribution is one of the most important problems in the theory of Markov chains. A simple answer can be given in the transient case (cf. example 2.20):

Theorem 2.22 *A transient Markov chain (i.e. a Markov chain with transient states only) has no stationary distribution.*

Proof: Assume that $\pi P = \pi$ holds for some distribution π and take any enumeration $E = (s_n : n \in \mathbb{N})$ of the state space E. Choose any index $m \in \mathbb{N}$ with $\pi_{s_m} > 0$. Since $\sum_{n=1}^{\infty} \pi_{s_n} = 1$ is bounded, there is an index $M > m$ such that $\sum_{n=M}^{\infty} \pi_{s_n} < \pi_{s_m}$. Set $\varepsilon := \pi_{s_m} - \sum_{n=M}^{\infty} \pi_{s_n}$. According to theorem 2.17, there is an index $N \in \mathbb{N}$ such that $P^n(s_i, s_m) < \varepsilon$ for all $i \leq M$ and $n \geq N$. Then the stationarity of π implies

$$\pi_{s_m} = \sum_{i=1}^{\infty} \pi_{s_i} P^N(s_i, s_m) = \sum_{i=1}^{M-1} \pi_{s_i} P^N(s_i, s_m) + \sum_{i=M}^{\infty} \pi_{s_i} P^N(s_i, s_m)$$

$$< \varepsilon + \sum_{i=M}^{\infty} \pi_{s_i} = \pi_{s_m}$$

which is a contradiction.
□

For the recurrent case, a finer distinction will be necessary. While the expected total number r_{jj} of visits to a recurrent state $j \in E$ is always infinite (see corollary 2.15), there are differences in the rate of visits to a recurrent state. In order to describe these, define $N_i(n)$ as the number of visits to state i until time n. Further define for a recurrent state $i \in E$ the mean time

$$m_i := \mathbb{E}(\tau_i | X_0 = i)$$

until the first visit to i (after time zero) under the condition that the chain starts in i. By definition $m_i > 0$ for all $i \in E$. The elementary renewal theorem (which will be proven later as theorem 6.12) states that

$$\lim_{n\to\infty} \frac{\mathbb{E}(N_i(n)|X_0 = j)}{n} = \frac{1}{m_i} \qquad (2.8)$$

for all recurrent $i \in E$ and independently of $j \in E$ provided $j \leftrightarrow i$, with the convention of $1/\infty := 0$. Thus the asymptotic rate of visits to a recurrent state is determined by the mean recurrence time of this state. This gives reason to the following definition: A recurrent state $i \in E$ with $m_i = \mathbb{E}(\tau_i|X_0 = i) < \infty$ will be called **positive recurrent**, otherwise i is called **null recurrent**. The distinction between positive and null recurrence is supported by the equivalence relation $\leftrightarrow$, as shown in

Theorem 2.23 *Positive recurrence and null recurrence are class properties with respect to the relation of communication between states.*

Proof: Assume that $i \leftrightarrow j$ for two states $i, j \in E$ and i is null recurrent. Thus there are numbers $m, n \in \mathbb{N}$ with $P^n(i, j) > 0$ and $P^m(j, i) > 0$. Because of the representation $\mathbb{E}(N_i(k)|X_0 = i) = \sum_{l=0}^{k} P^l(i, i)$, we obtain

$$
\begin{aligned}
0 &= \lim_{k\to\infty} \frac{\sum_{l=0}^{k} P^l(i, i)}{k} \\
&\geq \lim_{k\to\infty} \frac{\sum_{l=0}^{k-m-n} P^l(j, j)}{k} \cdot P^n(i, j) P^m(j, i) \\
&= \lim_{k\to\infty} \frac{k - m - n}{k} \cdot \frac{\sum_{l=0}^{k-m-n} P^l(j, j)}{k - m - n} \cdot P^n(i, j) P^m(j, i) \\
&= \lim_{k\to\infty} \frac{\sum_{l=0}^{k} P^l(j, j)}{k} \cdot P^n(i, j) P^m(j, i) \\
&= \frac{P^n(i, j) P^m(j, i)}{m_j}
\end{aligned}
$$

and thus $m_j = \infty$, which signifies the null recurrence of j.
$\square$

Thus we can call a communication class positive recurrent or null recurrent. In the former case, a construction of a stationary distribution is given in

Theorem 2.24 *Let $i \in E$ be positive recurrent and define the mean first visit time $m_i := \mathbb{E}(\tau_i|X_0 = i)$. Then a stationary distribution π is given by*

$$\pi_j := m_i^{-1} \cdot \sum_{n=0}^{\infty} \mathbb{P}(X_n = j, \tau_i > n|X_0 = i)$$

for all $j \in E$. In particular, $\pi_i = m_i^{-1}$ and $\pi_k = 0$ for all states k outside of the communication class belonging to i.

Proof: First of all, π is a probability measure since

$$\sum_{j \in E} \sum_{n=0}^{\infty} \mathbb{P}(X_n = j, \tau_i > n | X_0 = i) = \sum_{n=0}^{\infty} \sum_{j \in E} \mathbb{P}(X_n = j, \tau_i > n | X_0 = i)$$

$$= \sum_{n=0}^{\infty} \mathbb{P}(\tau_i > n | X_0 = i) = m_i$$

The particular statements in the theorem are obvious from theorem 2.16 and the definition of π. The stationarity of π is shown as follows. First we obtain

$$\pi_j = m_i^{-1} \cdot \sum_{n=0}^{\infty} \mathbb{P}(X_n = j, \tau_i > n | X_0 = i)$$

$$= m_i^{-1} \cdot \sum_{n=1}^{\infty} \mathbb{P}(X_n = j, \tau_i \geq n | X_0 = i)$$

$$= m_i^{-1} \cdot \sum_{n=1}^{\infty} \mathbb{P}(X_n = j, \tau_i > n - 1 | X_0 = i)$$

since $X_0 = X_{\tau_i} = i$ in the conditioning set $\{X_0 = i\}$. Because of

$$\mathbb{P}(X_n = j, \tau_i > n - 1 | X_0 = i)$$

$$= \frac{\mathbb{P}(X_n = j, \tau_i > n - 1, X_0 = i)}{\mathbb{P}(X_0 = i)}$$

$$= \sum_{k \in E} \frac{\mathbb{P}(X_n = j, X_{n-1} = k, \tau_i > n - 1, X_0 = i)}{\mathbb{P}(X_0 = i)}$$

$$= \sum_{k \in E \setminus \{i\}} \frac{\mathbb{P}(X_n = j, X_{n-1} = k, \tau_i > n - 1, X_0 = i)}{\mathbb{P}(X_{n-1} = k, \tau_i > n - 1, X_0 = i)}$$

$$\times \frac{\mathbb{P}(X_{n-1} = k, \tau_i > n - 1, X_0 = i)}{\mathbb{P}(X_0 = i)}$$

$$= \sum_{k \in E} p_{kj} \mathbb{P}(X_{n-1} = k, \tau_i > n - 1 | X_0 = i)$$

we can transform further

$$\pi_j = m_i^{-1} \cdot \sum_{n=1}^{\infty} \sum_{k \in E} p_{kj} \mathbb{P}(X_{n-1} = k, \tau_i > n - 1 | X_0 = i)$$

$$= \sum_{k \in E} p_{kj} \cdot m_i^{-1} \sum_{n=0}^{\infty} \mathbb{P}(X_n = k, \tau_i > n | X_0 = i) = \sum_{k \in E} \pi_k p_{kj}$$

which completes the proof.

□

Theorem 2.25 *Let $\mathcal{X}$ denote an irreducible, positive recurrent Markov chain. Then $\mathcal{X}$ has a unique stationary distribution.*

Proof: Existence has been shown in theorem 2.24. Uniqueness of the stationary distribution can be seen as follows. Let π denote the stationary distribution as constructed in theorem 2.24 and i the positive recurrent state that served as recurrence point for π. Further, let ν denote any stationary distribution for $\mathcal{X}$. Then there is a state $j \in E$ with $\nu_j > 0$ and a number $m \in \mathbb{N}$ with $P^m(j, i) > 0$, since $\mathcal{X}$ is irreducible. Consequently we obtain

$$\nu_i = \sum_{k \in E} \nu_k P^m(k, i) \geq \nu_j P^m(j, i) > 0$$

Hence we can multiply ν by a skalar factor c such that $c \cdot \nu_i = \pi_i = 1/m_i$. Denote $\tilde{\nu} := c \cdot \nu$.

Let $\tilde{P}$ denote the transition matrix P without the ith column, i.e. we define the (j, k)th entry of $\tilde{P}$ by $\tilde{p}_{jk} = p_{jk}$ if $k \neq i$ and zero otherwise. Denote further the Dirac measure on i by δ^i, i.e. $\delta_j^i = 1$ if $i = j$ and zero otherwise. Then the stationary distribution π can be represented by $\pi = m_i^{-1} \cdot \delta^i \sum_{n=0}^{\infty} \tilde{P}^n$.

We first claim that $m_i \tilde{\nu} = \delta^i + m_i \tilde{\nu} \tilde{P}$. This is clear for the entry $\tilde{\nu}_i$ and easily seen for $\tilde{\nu}_j$ with $j \neq i$ because in this case $(\tilde{\nu} \tilde{P})_j = c \cdot (\nu P)_j = \tilde{\nu}_j$. Now we can proceed with the same argument to see that

$$m_i \tilde{\nu} = \delta^i + (\delta^i + m_i \tilde{\nu} \tilde{P}) \tilde{P} = \delta^i + \delta^i \tilde{P} + m_i \tilde{\nu} \tilde{P}^2 = \dots$$

$$= \delta^i \sum_{n=0}^{\infty} \tilde{P}^n = m_i \pi$$

Hence $\tilde{\nu}$ already is a probability measure and the skalar factor must be $c = 1$. This yields $\nu = \tilde{\nu} = \pi$ and thus the statement.

□

Remark 2.26 At a closer look the assumption of irreducibility may be relaxed to some extend. For example, if there is exactly one closed positive recurrent communication class and a set of transient and inaccessible states (i.e. states j

for which there is no state i with $i \to j$), then the above statement still holds although $\mathcal{X}$ is not irreducible.

A first consequence of the uniqueness is the following simpler representation of the stationary distribution:

Theorem 2.27 *Let $\mathcal{X}$ denote an irreducible, positive recurrent Markov chain. Then the stationary distribution π of $\mathcal{X}$ is given by*

$$\pi_j = m_j^{-1} = \frac{1}{\mathbb{E}(\tau_j | X_0 = j)}$$

for all $j \in E$.

Proof: Since all states in E are positive recurrent, the construction in theorem 2.24 can be pursued for any inital state j. This yields $\pi_j = m_j^{-1}$ for all $j \in E$. The statement now follows from the uniqueness of the stationary distribution. □

Corollary 2.28 *For an irreducible, positive recurrent Markov chain, the stationary probability π_j of a state j coincides with its asymptotic rate of recurrence, i.e.*

$$\lim_{n \to \infty} \frac{\mathbb{E}(N_j(n) | X_0 = i)}{n} = \pi_j$$

for all $j \in E$ and independently of $i \in E$. Further, if an asymptotic distribution $p = \lim_{n \to \infty} \mathbb{P}(X_n = .)$ does exist, then it coincides with the stationary distribution. In particular, it is independent of the initial distribution of $\mathcal{X}$.

Proof: The first statement immediately follows from equation (2.8). For the second statement, it suffices to employ $\mathbb{E}(N_j(n) | X_0 = i) = \sum_{l=0}^{n} P^l(i, j)$. If an asymptotic distribution p does exist, then for any initial distribution ν we obtain

$$p_j = \lim_{n \to \infty} (\nu P^n)_j = \sum_{i \in E} \nu_i \lim_{n \to \infty} P^n(i, j)$$

$$= \sum_{i \in E} \nu_i \lim_{n \to \infty} \frac{\sum_{l=0}^{n} P^l(i, j)}{n} = \sum_{i \in E} \nu_i \pi_j$$

$$= \pi_j$$

independently of ν.
□

4.   Restricted Markov Chains

Now let $F \subset E$ denote any subset of the state space E. Define $\tau_F(k)$ to be the stopping time of the kth visit of $\mathcal{X}$ to the set F, i.e.

$$\tau_F(k+1) := \min\{n > \tau_F(k) : X_n \in F\}$$

with $\tau_F(0) := 0$. If $\mathcal{X}$ is recurrent, then the strong Markov property (theorem 2.7) ensures that the chain $\mathcal{X}^F = (X_n^F : n \in \mathbb{N})$ with $X_n^F := X_{\tau_F(n)}$ is a recurrent Markov chain, too. It is called the Markov chain restricted to F. In case of positive recurrence, we can obtain the stationary distribution of $\mathcal{X}^F$ from the stationary distribution of $\mathcal{X}$ in a simple manner:

Theorem 2.29 *If the Markov chain $\mathcal{X}$ is positive recurrent, then the stationary distribution of $\mathcal{X}^F$ is given by*

$$\pi_j^F = \frac{\pi_j}{\sum_{k \in F} \pi_k}$$

for all $j \in F$.

Proof: Choose any state $i \in F$ and recall from theorem 2.24 the expression

$$\pi_j := m_i^{-1} \cdot \sum_{n=0}^{\infty} \mathbb{P}(X_n = j, \tau_i > n | X_0 = i)$$

which holds for all $j \in F$. For π_j^F we can perform the same construction with respect to the chain $\mathcal{X}^F$. By the definition of $\mathcal{X}^F$ it is clear that the number of visits to the state j between two consecutive visits to i is the same for the chains $\mathcal{X}$ and $\mathcal{X}^F$. Hence the sum expression for π_j^F, which is the expectation of that number of visits, remains the same as for π_j. The other factor m_i^{-1} in the formula above is independent of j and serves only as a normalization constant, i.e. in order to secure that $\sum_{j \in E} \pi_j = 1$. Hence for a construction of π_j^F with respect to $\mathcal{X}^F$ this needs to be replaced by $(m_i \cdot \sum_{k \in F} \pi_k)^{-1}$, which then yields the statement.
□

Theorem 2.30 *Let $\mathcal{X} = (X_n : n \in \mathbb{N}_0)$ denote an irreducible and positive recurrent Markov chain with discrete state space E. Further let $F \subset E$ denote any subset of E, and $\mathcal{X}^F$ the Markov chain restricted to F. Denote*

$$\tau_F := \min\{n \in \mathbb{N} : X_n \in F\}$$

Then a measure ν on E is stationary for $\mathcal{X}$ if and only if $\nu' = (\nu_i : i \in F)$ is stationary for $\mathcal{X}^F$ and

$$\nu_j = \sum_{k \in F} \nu_k \sum_{n=0}^{\infty} \mathbb{P}(X_n = j, \tau_F > n | X_0 = k) \tag{2.9}$$

for all $j \in E \setminus F$.

Proof: Due to theorem 2.29 it suffices to prove equation (2.9) for $j \in E \setminus F$. Choose any state $i \in F$ and define

$$\tau_i := \min\{n \in \mathbb{N} : X_n = i\}$$

According to theorem 2.24 the stationary measure v for $\mathcal{X}$ is given by

$$\nu_j = \nu_i \cdot \sum_{n=0}^{\infty} \mathbb{P}(X_n = j, \tau_i > n | X_0 = i) = \nu_i \cdot \mathbb{E}_i \left(\sum_{n=0}^{\tau_i - 1} \mathbf{1}_{X_n = j} \right)$$

for $j \in E \setminus F$, where $\mathbb{E}_i$ denotes the conditional expectation given $X_0 = i$. Define further

$$\tau_i^F := \min\{n \in \mathbb{N} : X_n^F = i\}$$

Because of the strong Markov property we can proceed as

$$\nu_j = \nu_i \cdot \mathbb{E}_i \left(\sum_{n=0}^{\tau_i^F - 1} \mathbb{E}_{X_n^F} \sum_{m=0}^{\tau_F - 1} \mathbf{1}_{X_m = j} \right)$$

$$= \nu_i \cdot \sum_{k \in F} \mathbb{E}_i \left(\sum_{n=0}^{\tau_i^F - 1} \mathbf{1}_{X_n^F = k} \right) \cdot \mathbb{E}_k \left(\sum_{m=0}^{\tau_F - 1} \mathbf{1}_{X_m = j} \right)$$

Regarding the restricted Markov chain $\mathcal{X}^F$, theorem 2.24 states that

$$\mathbb{E}_i \left(\sum_{n=0}^{\tau_i^F - 1} \mathbf{1}_{X_n^F = k} \right) = \sum_{n=0}^{\infty} \mathbb{P}(X_n^F = k, \tau_i^F > n | X_0^F = i) = \frac{\nu_k}{\nu_i}$$

for all $k \in F$. Hence we obtain

$$\nu_j = \sum_{k \in F} \nu_k \sum_{n=0}^{\infty} \mathbb{P}(X_n = j, \tau_F > n | X_0 = k)$$

which was to be proven.

$\square$

5. Conditions for Positive Recurrence

In the third part of this course we will need some results on the behaviour of a Markov chain on a finite subset of its state space. As a first fundamental result we state

Theorem 2.31 *An irreducible Markov chain with finite state space F is positive recurrent.*

Proof: For all $n \in \mathbb{N}$ and $i \in F$ we have $\sum_{j \in E} P^n(i, j) = 1$. Hence it is not possible that $\lim_{n \to \infty} P^n(i, j) = 0$ for all $j \in F$. Thus there is one state $h \in F$ such that $r_{hh} = \sum_{n=0}^{\infty} P^n(h, h) = \infty$, which means by corollary 2.15 that h is recurrent and by irreducibility that the chain is recurrent.

If the chain were null recurrent, then according to the relation in (2.8)

$$\lim_{n \to \infty} \frac{1}{n} \sum_{k=1}^{n} P^k(i, j) = 0$$

would hold for all $j \in F$, independently of i because of irreducibility. But this would imply that $\lim_{n \to \infty} P^n(i, j) = 0$ for all $j \in F$, which contradicts our first observation in this proof. Hence the chain must be positive recurrent.
□

For irreducible Markov chains the condition $\mathbb{E}(\tau_i | X_0 = i) < \infty$ implies positive recurrence of state i and hence positive recurrence of the whole chain. Writing τ_F for the time of the first visit to the set F, we now can state the following generalization of this condition:

Theorem 2.32 *Let $\mathcal{X}$ denote an irreducible Markov chain with state space E and be $F \subset E$ a finite subset of E. The chain $\mathcal{X}$ is positive recurrent if and only if $\mathbb{E}(\tau_F | X_0 = i) < \infty$ for all $i \in F$.*

Proof: If $\mathcal{X}$ is positive recurrent, then $\mathbb{E}(\tau_F | X_0 = i) \leq \mathbb{E}(\tau_i | X_0 = i) < \infty$ for all $i \in F$, by the definition of positive recurrence.

Now assume that $\mathbb{E}(\tau_F | X_0 = i) < \infty$ for all $i \in F$. Define the stopping times $\sigma(i) := \min\{k \in \mathbb{N} : X_k^F = i\}$ and random variables $Y_k := \tau_F(k) - \tau_F(k-1)$. Since F is finite, $m := \max_{j \in F} \mathbb{E}(\tau_F | X_0 = j) < \infty$. We shall denote the conditional expectation given $X_0 = i$ by $\mathbb{E}_i$. For $i \in F$ we now obtain

$$\mathbb{E}(\tau_i | X_0 = i) = \mathbb{E}_i \left(\sum_{k=1}^{\sigma(i)} Y_k \right) = \sum_{k=1}^{\infty} \mathbb{E}_i \left(\mathbb{E}(Y_k | X_{\tau_F(k-1)}) \cdot 1_{k \leq \sigma(i)} \right)$$

$$\leq m \cdot \sum_{k=1}^{\infty} \mathbb{P}(\sigma(i) \geq k | X_0 = i) = m \cdot \mathbb{E}(\sigma(i) | X_0 = i)$$

Since F is finite, $\mathcal{X}^F$ is positive recurrent by theorem 2.31. Hence we know that $\mathbb{E}(\sigma(i) | X_0 = i) < \infty$, and thus $\mathbb{E}(\tau_i | X_0 = i) < \infty$ which shows that $\mathcal{X}$ is positive recurrent.
□

An often difficult problem is to determine whether a given Markov chain is positive recurrent or not. Concerning this, we now introduce one of the most important criteria for the existence of stationary distributions of Markov chains occuring in queueing theory. It is known as **Foster's criterion**.

Theorem 2.33 *Let $\mathcal{X}$ denote an irreducible Markov chain with countable state space E and transition matrix P. Further let F denote a finite subset of E. If there is a function $h : E \to \mathbb{R}$ with $\inf\{h(i) : i \in E\} > -\infty$, such that the conditions*

$$\sum_{k \in E} p_{ik} h(k) < \infty \qquad \text{and} \qquad \sum_{k \in E} p_{jk} h(k) \leq h(j) - \varepsilon$$

hold for some $\varepsilon > 0$ and all $i \in F$ and $j \in E \setminus F$, then $\mathcal{X}$ is positive recurrent.

Proof: Without loss of generality we can assume $h(i) \geq 0$ for all $i \in E$, since otherwise we only need to increase h by a suitable constant. Define the stopping time $\tau_F := \min\{n \in \mathbb{N}_0 : X_n \in F\}$. First we observe that

$$\mathbb{E}(h(X_{n+1}) \cdot 1_{\tau_F > n+1} | X_0, \ldots, X_n) \leq \mathbb{E}(h(X_{n+1}) \cdot 1_{\tau_F > n} | X_0, \ldots, X_n)$$

$$= 1_{\tau_F > n} \cdot \sum_{k \in E} p_{X_n, k} h(k)$$

$$\leq 1_{\tau_F > n} \cdot (h(X_n) - \varepsilon)$$

$$= h(X_n) \cdot 1_{\tau_F > n} - \varepsilon \cdot 1_{\tau_F > n}$$

holds for all $n \in \mathbb{N}_0$, where the first equality is due to (15.3). We now proceed with

$$
\begin{aligned}
0 &\leq \mathbb{E}(h(X_{n+1}) \cdot 1_{\tau_F > n+1} | X_0 = i) \\
&= \mathbb{E}(\mathbb{E}(h(X_{n+1}) \cdot 1_{\tau_F > n+1} | X_0, \ldots, X_n) | X_0 = i) \\
&\leq \mathbb{E}(h(X_n) \cdot 1_{\tau_F > n} | X_0 = i) - \varepsilon \mathbb{P}(\tau_F > n | X_0 = i) \\
&\leq \ldots \\
&\leq \mathbb{E}(h(X_0) \cdot 1_{\tau_F > 0} | X_0 = i) - \varepsilon \sum_{k=0}^{n} \cdot \mathbb{P}(\tau_F > k | X_0 = i)
\end{aligned}
$$

which holds for all $i \in E \setminus F$ and $n \in \mathbb{N}_0$. For $n \to \infty$ this implies

$$
\mathbb{E}(\tau_F | X_0 = i) = \sum_{k=0}^{\infty} \mathbb{P}(\tau_F > k | X_0 = i) \leq h(i)/\varepsilon < \infty
$$

for $i \in E \setminus F$. Now the mean return time to the state set F is bounded by

$$
\begin{aligned}
\mathbb{E}(\tau_F | X_0 = i) &= \sum_{j \in F} p_{ij} + \sum_{j \in E \setminus F} p_{ij} \mathbb{E}(\tau_F + 1 | X_0 = j) \\
&\leq 1 + \varepsilon^{-1} \sum_{j \in E} p_{ij} h(j) < \infty
\end{aligned}
$$

for all $i \in F$, which completes the proof. $\square$

6. The M/M/1 queue in discrete time

Choose any parameters $0 < p, q < 1$. Let the arrival process be distributed as a Bernoulli process with parameter p and the service times $(S_n : n \in \mathbb{N}_0)$ be iid according to the geometric distribution with parameter q.

The geometric service time distribution and the Bernoulli arrival process have been chosen because this simplifies the formulation of the system process in terms of a Markov model due to the following **memoryless property**:

Theorem 2.34 *Let S be distributed geometrically with parameter q, i.e. let $\mathbb{P}(S = k) = (1 - q)^{k-1} q$ for all $k \in \mathbb{N}$. Then $\mathbb{P}(S = k | S > k - 1) = q$ holds for the conditional distribution, independently of k. Likewise, if Z_n is the nth inter–arrival time of a Bernoulli process with parameter p, then the relation $\mathbb{P}(Z_n = k | Z_n > k - 1) = p$ holds, independently of k and n.*

Proof: First the proof for the geometric distribution: For all $k \in \mathbb{N}$, the argument

$$\mathbb{P}(S = k | S > k - 1) = \frac{\mathbb{P}(S = k, S > k - 1)}{\mathbb{P}(S > k - 1)} = \frac{\mathbb{P}(S = k)}{\mathbb{P}(S > k - 1)}$$

$$= \frac{(1 - q)^{k-1} q}{(1 - q)^{k-1}} = q$$

holds, which shows the first statement. For a Bernoulli process, the nth inter–arrival time $Z_n = T_n - T_{n-1}$ is distributed geometrically with parameter p, due to the strong Markov property. This completes the proof for the second statement.

□

Thus the memoryless property states that no matter how long a service time or an inter–arrival time has already passed, the probability of a service completion or an arrival at the next time instant is always the same. Hence the system process $\mathcal{Q} = (Q_n : n \in \mathbb{N}_0)$ of the M/M/1 queue in discrete time with arrival process $\mathcal{T}$ and service times S_n can be formulated easily as a homogeneous Markov chain. It has state space $E = \mathbb{N}_0$ and transition probabilities $p_{01} := p$, $p_{00} := 1 - p$, and

$$p_{ij} := \begin{cases} p(1 - q), & j = i + 1 \\ pq + (1 - p)(1 - q), & j = i \\ q(1 - p), & j = i - 1 \end{cases}$$

for $i \geq 1$. Because of the simple state space, the transition matrix can be displayed in the form of a triagonal matrix

$$P = \begin{pmatrix} 1 - p & p & 0 & \cdots \\ q(1 - p) & pq + (1 - p)(1 - q) & p(1 - q) & \ddots \\ 0 & q(1 - p) & pq + (1 - p)(1 - q) & \ddots \\ \vdots & \ddots & \ddots & \ddots \end{pmatrix}$$

Since $p, q > 0$, the chain $\mathcal{Q}$ is irreducible. If $p < q$, then $h(n) := n$ defines a function which satisfies the conditions for Foster's criterion, as

$$\sum_{k=0}^{\infty} p_{ik} h(k) = q(1 - p) \cdot (i - 1) + (qp + (1 - q)(1 - p)) \cdot i$$

$$+ p(1 - q) \cdot (i + 1)$$

$$= i - q(1 - p) + p(1 - q) = i - q + p \leq i - \varepsilon$$

for all $i \in \mathbb{N}$, with $\varepsilon = q - p > 0$, and $\sum_{k=0}^{\infty} p_{0k} \cdot h(k) = p < \infty$ show. The ratio p/q is called the **load** of the queue. Thus the system process $\mathcal{Q}$ is positive recurrent if the queue load is less than one.

In order to derive a stationary distribution for $\mathcal{Q}$, we first introduce notations $p' := p(1 - q)$ and $q' := q(1 - p)$. Then we translate the condition $\pi P = \pi$ into the equations

$$\pi_0 = \pi_0(1 - p) + \pi_1 q' \tag{2.10}$$

$$\pi_1 = \pi_0 p + \pi_1(1 - p' - q') + \pi_2 q' \tag{2.11}$$

$$\pi_n = \pi_{n-1}p' + \pi_n(1 - (p' + q')) + \pi_{n+1}q' \tag{2.12}$$

for all $n \geq 2$. For the solution, we guess the geometric form

$$\pi_{n+1} = \pi_n \cdot r$$

for all $n \geq 1$, with $r > 0$. Thus equation (2.12) becomes

$$0 = \pi_n p' - \pi_n r(p' + q') + \pi_n r^2 q' = \pi_n \left(p' - r(p' + q') + r^2 q' \right)$$

for all $n \geq 1$, which leads for non–trivial $\pi \neq 0$ to the roots $r = 1$ and $r = p'/q'$ of the quadratic term.

In the first case $r = 1$, we obtain $\pi_{n+1} = \pi_n$ for all $n \geq 1$. This implies $\sum_{j \in E} \pi_j = \infty$ and thus cannot lead to a stationary distribution. Hence in the case $r = 1$ the geometric approach is not successful.

The second root $r = p'/q'$ allows solutions for the other equations (2.10) and (2.11) too. This can be checked as follows: First, the relation

$$\pi_1 = \pi_0 \frac{p}{q'} = \pi_0 \frac{\rho}{1 - p}$$

is a requirement from equation (2.10). Then the second equation (2.11) yields

$$\pi_2 = \frac{1}{q'} \left(\pi_1(p' + q') - \pi_0 p \right) = \frac{1}{q'} \left(\frac{p}{q'}(p' + q') - p \right) \pi_0$$

$$= \pi_0 \frac{p}{q'} \left(\frac{p' + q'}{q'} - 1 \right) = \pi_1 \frac{p'}{q'}$$

in accordance with our geometric approach. Now normalization of π leads to

$$1 = \sum_{n=0}^{\infty} \pi_n = \pi_0 \left(1 + \frac{p}{q'} \sum_{n=1}^{\infty} \left(\frac{p'}{q'} \right)^{n-1} \right)$$

from which we obtain

$$
\pi_0 = \left(1 + \frac{p}{q'} \sum_{n=1}^{\infty} \left(\frac{p'}{q'}\right)^{n-1}\right)^{-1} = \left(1 + \frac{p}{q'(1 - p'/q')}\right)^{-1}
$$

$$
= \left(1 + \frac{p}{q' - p'}\right)^{-1} = (q' - p')(q' - p' + p)^{-1} = \frac{q - p}{q}
$$

$$
= 1 - \rho
$$

with $\rho := p/q$, because of $q' - p' = q - p$. Hence the approach $\pi_{n+1} = \pi_n \cdot r$ with $r = p'/q'$ leads to a solution of $\pi P = \pi$.

Note that $r < 1$ if and only if $p < q$. Further, the mean inter–arrival time is $\mathbb{E}(T_1) = 1/p$ and the mean service time is $\mathbb{E}(S_1) = 1/q$. Thus the geometric approach is successful if the so–called **stability condition**

$$
\rho = \frac{p}{q} = \frac{\mathbb{E}(S_1)}{\mathbb{E}(T_1)} < 1
$$

holds. This condition simply postulates that the mean service time be shorter than the mean inter–arrival time. In this case, the stationary distribution π of $\mathcal{Q}$ has the form

$$
\pi_0 = 1 - \rho \qquad \text{and} \qquad \pi_n = (1 - \rho)\frac{\rho}{1 - p} r^{n-1}
$$

for all $n \geq 1$. It thus is a modified geometric distribution with parameter $r = p'/q' < 1$.

Notes

Markov chains originate from a series of papers written by A. Markov at the beginning of the 20th century. His first application is given here as exercise 2.3. However, methods and terminology at that time were very different from today's presentations.

The literature on Markov chains is perhaps the most extensive in the field of stochastic processes. This is not surprising, as Markov chains form a simple and useful starting point for the introduction of other processes.

Textbook presentations are given in Feller [34], Breiman [16], Karlin and Taylor [46], or Çinlar [25], to name but a few. The treatment in Ross [75] contains the useful concept of time–reversible Markov chains. An exhaustive introduction to Markov chains on general state spaces and conditions for their positive recurrence is given in Meyn and Tweedie [59].

Exercise 2.1 Let $(X_n : n \in \mathbb{N}_0)$ be a family of iid random variables with discrete state space. Show that $\mathcal{X} = (X_n : n \in \mathbb{N}_0)$ is a homogeneous Markov chain.

Exercise 2.2 Let $(X_n : n \in \mathbb{N}_0)$ be iid random variables on $\mathbb{N}_0$ with probabilities $a_i := \mathbb{P}(X_n = i)$ for all $n, i \in \mathbb{N}_0$. The event $X_n > \max(X_0, \ldots, X_{n-1})$ for $n \geq 1$ is called a record at time n. Define T_i as the time of the ith record, i.e. $T_0 := 0$ and $T_{i+1} := \min\{n \in \mathbb{N} : X_n > X_{T_i}\}$ for all $i \in \mathbb{N}_0$. Denote the ith record value by $R_i := X_{T_i}$. Show that $(R_i : i \in \mathbb{N}_0)$ and $((R_i, T_i) : i \in \mathbb{N}_0)$ are Markov chains by determining their transition probabilities.

Exercise 2.3 Diffusion model by Bernoulli and Laplace
The following is a stochastic model for the flow of two incompressible fluids between two containers: Two boxes contain m balls each. Of these $2m$ balls, b are black and the others are white. The system is said to be in state i if the first box contains i black balls. A state transition is performed by choosing one ball out of each box at random (meaning here that each ball is chosen with equal probability) and then interchanging the two. Derive a Markov chain model for the system and determine the transition probabilities.

Exercise 2.4 Let $\mathcal{X}$ denote a Markov chain with $m < \infty$ states. Show that if state j is accessible from state i, then it is accessible in at most $m - 1$ transitions.

Exercise 2.5 Let $p = (p_n : n \in \mathbb{N}_0)$ be a discrete probability distribution and define

$$P = \begin{pmatrix} p_0 & p_1 & p_2 & \cdots \\ & p_0 & p_1 & \ddots \\ & & p_0 & \ddots \\ & & & \ddots \end{pmatrix}$$

with all non–specified entries being zero. Let $\mathcal{X}$ denote a Markov chain with state space $\mathbb{N}_0$ and transition matrix P. Derive an expression (in terms of discrete convolutions) for the transition probabilities $\mathbb{P}(X_{n+m} = j | X_n = i)$ with $n, m \in \mathbb{N}_0$ and $i, j \in \mathbb{N}_0$. Apply the result to the special case of a Bernoulli process (see example 2.3).

Exercise 2.6 Prove equation (2.6).

Exercise 2.7 Prove the equation $P^n(i, j) = \sum_{k=1}^{n} F_k(i, j) P^{n-k}(j, j)$ for all $n \in \mathbb{N}$ and $i, j \in E$.

Exercise 2.8 Let $\mathcal{X}$ denote a Markov chain with state space $E = \{1, \ldots, 10\}$ and transition matrix

$$P = \begin{pmatrix} 1/2 & 0 & 1/2 & 0 & 0 & 0 & 0 & 0 & 0 & 0 \\ 0 & 1/3 & 0 & 0 & 0 & 0 & 2/3 & 0 & 0 & 0 \\ 1 & 0 & 0 & 0 & 0 & 0 & 0 & 0 & 0 & 0 \\ 0 & 0 & 0 & 0 & 1 & 0 & 0 & 0 & 0 & 0 \\ 0 & 0 & 0 & 1/3 & 1/3 & 0 & 0 & 0 & 1/3 & 0 \\ 0 & 0 & 0 & 0 & 0 & 1 & 0 & 0 & 0 & 0 \\ 0 & 0 & 0 & 0 & 0 & 0 & 1/4 & 0 & 3/4 & 0 \\ 0 & 0 & 1/4 & 1/4 & 0 & 0 & 0 & 1/4 & 0 & 1/4 \\ 0 & 1 & 0 & 0 & 0 & 0 & 0 & 0 & 0 & 0 \\ 0 & 1/3 & 0 & 0 & 1/3 & 0 & 0 & 0 & 0 & 1/3 \end{pmatrix}$$

Reorder the states according to their communication classes and determine the resulting form of the transition matrix as in representation (2.4). Determine further a transition graph, in which

means that $f_{ij} > 0$.

Exercise 2.9 Prove equation (2.7).
Hint: Derive a representation of N_j in terms of the random variables

$$A_n := \begin{cases} 1, & X_n = j \\ 0, & X_n \neq j \end{cases}$$

Exercise 2.10 Prove corollary 2.15.

Exercise 2.11 Prove remark 2.26.

Exercise 2.12 A server's up time is k time units with probability $p_k = 2^{-k}$, $k \in \mathbb{N}$. After failure the server is immediately replaced by an identical new one. The up time of the new server is of course independent of the behaviour of all preceding servers.
Let X_n denote the remaining up time of the server at time $n \in \mathbb{N}_0$. Determine the transition probabilities for the Markov chain $\mathcal{X} = (X_n : n \in \mathbb{N}_0)$ and determine the stationary distribution of $\mathcal{X}$.

Exercise 2.13 Let P denote the transition matrix of an irreducible Markov chain $\mathcal{X}$ with discrete state space $E = F \cup F^c$, where $F^c = E \setminus F$. Write P in block notation as

$$P = \begin{pmatrix} P_{FF} & P_{FF^c} \\ P_{F^cF} & P_{F^cF^c} \end{pmatrix}$$

Show that the Markov chain $\mathcal{X}^F$ restricted to the state space F has transition matrix

$$P^F = P_{FF} + P_{FF^c}(I - P_{F^cF^c})^{-1}P_{F^cF}$$

with I denoting the identity matrix on F^c.

Exercise 2.14 Let $\mathcal{X}$ denote a Markov chain with state space $E = \{0, \dots, m\}$ and transition matrix

$$P = \begin{pmatrix} p_{00} & p_{01} & & & \\ p_{10} & p_{11} & p_{12} & & \\ & p_{21} & p_{22} & p_{23} & \\ & & \ddots & \ddots & \ddots \\ & & & p_{m,m-1} & p_{mm} \end{pmatrix}$$

where $p_{ij} > 0$ for $|i - j| = 1$. Show that the stationary distribution π of $\mathcal{X}$ is uniquely determined by

$$\pi_n = \pi_0 \cdot \prod_{i=1}^{n} \frac{p_{i-1,i}}{p_{i,i-1}} \qquad \text{and} \qquad \pi_0 = \left(\sum_{j=0}^{m} \prod_{i=1}^{j} \frac{p_{i-1,i}}{p_{i,i-1}} \right)^{-1}$$

for all $n = 1, \dots, m$.
Use this result to determine the stationary distribution of the Bernoulli–Laplace diffusion model with $b = m$ (see exercise 2.3).

Exercise 2.15 Show that the second condition in theorem 2.33 can be substituted by the condition

$$\sum_{j \in E} p_{ij} h(j) \leq h(i) - 1 \qquad \text{for all } i \in E \setminus F.$$

Exercise 2.16 Show the following complement to theorem 2.33: Let P denote the transition matrix of a positive recurrent Markov chain with discrete state space E. Then there is a function $h : E \to \mathbb{R}$ and a finite subset $F \subset E$ such that

$$\sum_{j \in E} p_{ij} h(j) < \infty \qquad \text{for all } i \in F, \text{ and}$$

$$\sum_{j \in E} p_{ij} h(j) \leq h(i) - 1 \qquad \text{for all } i \in E \setminus F.$$

Hint: Consider the conditional expectation of the remaining time until return-ing to a fixed set F of states.

Exercise 2.17 For the discrete, non–negative random walk with transition ma-trix

$$P = \begin{pmatrix} p_{00} & p_{01} & & & \\ p_{10} & 0 & p_{12} & & \\ & p_{10} & 0 & p_{12} & \\ & & \ddots & \ddots & \ddots \end{pmatrix}$$

determine the criterion of positive recurrence according to theorem 2.33.

Chapter 3

HOMOGENEOUS MARKOV PROCESSES
ON DISCRETE STATE SPACES

In the present chapter we will transfer the discrete time results of the previous chapter to Markov processes in continuous time.

1. Definition

Define $T_0 := 0$ and let $(T_n : n \in \mathbb{N})$ denote a sequence of positive real–valued random variables with $T_{n+1} > T_n$ for all $n \in \mathbb{N}_0$ and $T_n \to \infty$ as $n \to \infty$. Further, let E denote a countable state space and $(X_n : n \in \mathbb{N}_0)$ a sequence of E–valued random variables. A process $\mathcal{Y} = (Y_t : t \in \mathbb{R}_0^+)$ in continuous time with

$$Y_t := X_n \qquad \text{for} \qquad T_n \leq t < T_{n+1}$$

is called a **pure jump process**. The variable $H_n := T_{n+1} - T_n$ (resp. X_n) is called the nth **holding time** (resp. the nth **state**) of the process $\mathcal{Y}$. If further $\mathcal{X} = (X_n : n \in \mathbb{N}_0)$ is a Markov chain with transition matrix $P = (p_{ij})_{i,j \in E}$ and the variables H_n are independent and distributed exponentially with parameter λ_{X_n} only depending on the state X_n, then $\mathcal{Y}$ is called homogeneous **Markov process** with discrete **state space** E. The chain $\mathcal{X}$ is called the **embedded Markov chain** of $\mathcal{Y}$. As a technical assumption we always agree upon the condition $\hat{\lambda} := \sup\{\lambda_i : i \in E\} < \infty$, i.e. the parameters for the exponential holding times shall be bounded.

An immediate consequence of the definition is that the paths of a Markov process are step functions. The lengths of the holding times are almost certainly strictly positive, since exponential distributions are zero with probability zero.

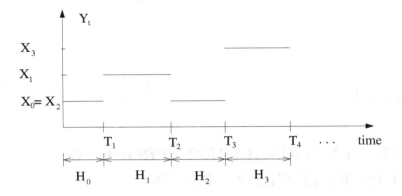

Figure 3.1. Typical path of a Markov process with discrete state space

Example 3.1 Poisson process

Define $X_n := n$ deterministically. Then $\mathcal{X} = (X_n : n \in \mathbb{N}_0)$ is a Markov chain with state space $E = \mathbb{N}_0$ and transition probabilities $p_{n,n+1} = 1$ for all $n \in \mathbb{N}_0$. Let the holding times H_n be distributed exponentially with identical parameter $\lambda > 0$. Then the resulting process $\mathcal{Y}$ as defined in the above definition is a Markov process with state space $\mathbb{N}_0$. It is called **Poisson process** with **intensity** (also: rate or parameter) λ.

Next we want to prove a property similar to the Markov property for Markov chains in discrete time. To this aim, we need to show the **memoryless property** for the exponential distribution, which is the analogue to the memoryless property for geometric distributions in discrete time.

Lemma 3.2 *Let H denote a random variable having an exponential distribution with parameter λ. Then the memoryless property*

$$\mathbb{P}(H > t + s | H > s) = \mathbb{P}(H > t)$$

holds for all time durations $s, t > 0$.

Proof: We immediately check

$$\mathbb{P}(H > t + s | H > s) = \frac{\mathbb{P}(H > t + s, H > s)}{\mathbb{P}(H > s)} = \frac{\mathbb{P}(H > t + s)}{\mathbb{P}(H > s)}$$

$$= \frac{e^{-\lambda \cdot (t+s)}}{e^{-\lambda \cdot s}} = e^{-\lambda \cdot t} = \mathbb{P}(H > t)$$

which holds for all $s, t > 0$.

□

Theorem 3.3 *Let $\mathcal{Y}$ denote a Markov process with discrete state space E. Then the* **Markov property**

$$\mathbb{P}(Y_t = j | Y_u : u \leq s) = \mathbb{P}(Y_t = j | Y_s)$$

holds for all times $s < t$ and states $j \in E$.

Proof: Denote the state at time s by $Y_s = i$. Because of the memoryless property of the exponential holding times, the remaining time in state i is distributed exponentially with parameter λ_i, no matter how long the preceeding holding time has been. After the holding time in the present state elapses, the process changes to another state j according to the homogeneous Markov chain $\mathcal{X}$. Hence the probability for the next state being j is given by p_{ij}, independently of any state of the process before time s. Now another exponential holding time begins, and thus the past before time s will not have any influence on the future of the process $\mathcal{Y}$.
$\square$

Analogous to the discrete time case, for any two time instances $s < t$ the conditional probabilities $\mathbb{P}(Y_t = j | Y_s = i)$ shall be called the **transition probabilities** from time s to time t. We will now derive a recursion formula for the transition probabilities of a Markov process by conditioning on the number of jumps between time s and time t:

Theorem 3.4 *The transition probabilities of a Markov process $\mathcal{Y}$ are given by*

$$\mathbb{P}(Y_t = j | Y_s = i) = \sum_{n=0}^{\infty} P_{ij}^{(n)}(s,t)$$

for all times $s < t$ and states $i, j \in E$, with

$$P_{ij}^{(0)}(s,t) = \delta_{ij} \cdot e^{-\lambda_i \cdot (t-s)}$$

and recursively

$$P_{ij}^{(n+1)}(s,t) = \int_s^t e^{-\lambda_i \cdot u} \lambda_i \sum_{k \in E} p_{ik} P_{kj}^{(n)}(u,t) \, du$$

for all $n \in \mathbb{N}_0$.

Proof: The above representation follows immediately by conditioning on the number of jumps in $]s,t]$. The expressions $P_{ij}^{(n)}(s,t)$ represent the conditional probabilities that $Y_t = j$ and there are n jumps in $]s,t]$ given that

$Y_s = i$. In the recursion formula the integral comprises all times u of a possible first jump along with the Lebesgue density $e^{-\lambda_i \cdot u} \lambda_i$ of this event, after which the probability of n remaining jumps reaching state j at time t is given by $\sum_{k \in E} p_{ik} P_{kj}^{(n)}(u, t)$.

$\square$

For every two time instances $s < t$, define the **transition probability matrix** $P(s, t)$ from time s to time t by its entries

$$P_{ij}(s, t) := \mathbb{P}(Y_t = j | Y_s = i)$$

Using the recursion formula, it is shown by induction on n that the conditional probabilities $P_{ij}^{(n)}(s, t)$ are homogeneous in time, i.e. they satisfy

$$P_{ij}^{(n)}(s, t) = P_{ij}^{(n)}(0, t - s)$$

for all $s < t$. Thus we can from now on restrict the analysis to the transition probability matrices

$$P(t) := P(0, t)$$

with $t \geq 0$. With this notation the Markov property yields the **Chapman–Kolmogorov equations**

$$P(s + t) = P(s)P(t)$$

for all time durations $s, t \geq 0$. Thus the family $\{P(t) : t \geq 0\}$ of transition probability matrices forms a semi–group under the composition of matrix multiplication. In particular, we obtain for the neutral element of this semi–group $P(0) = I_E := (\delta_{ij})_{i,j \in E}$ with $\delta_{ij} = 1$ for $i = j$ and zero otherwise.

In order to derive a simpler expression for the transition probability matrices, we need to introduce another concept, which will be called the **generator matrix**. This is defined as the matrix $G = (g_{ij})_{i,j \in E}$ on E with entries

$$g_{ij} := \begin{cases} -\lambda_i \cdot (1 - p_{ii}), & i = j \\ \lambda_i \cdot p_{ij}, & i \neq j \end{cases}$$

for all states $i, j \in E$. In particular, the relation

$$g_{ii} = -\sum_{j \neq i} g_{ij} \tag{3.1}$$

holds for all $i \in E$.

The (i, j)th entry of the generator G is called the **infinitesimal transition rate** from state i to state j. Using these, we can illustrate the dynamics of a Markov process in a directed graph where the nodes represent the states and an edge

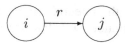

means that $g_{ij} = r > 0$. Such a graph is called a **state transition graph** of the Markov process. With the convention $p_{ii} = 0$ the state transition graph uniquely determines the Markov process.

Example 3.5 The state transition graph of the Poisson process with intensity λ (see example 3.1) is given by

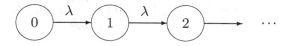

Figure 3.2. Poisson process

Theorem 3.6 *The transition probabilities $P_{ij}(t)$ of a Markov process satisfy the systems*

$$\frac{dP_{ij}(t)}{dt} = \sum_{k \in E} P_{ik}(t) g_{kj} = \sum_{k \in E} g_{ik} P_{kj}(t)$$

of differential equations. These are called the **Kolmogorov forward and backward equations**.

Proof: From the representation in theorem 3.4, it follows by induction on the number of jumps that all restricted probabilities $P^{(n)}(t)$ are Lebesgue integrable with respect to t over finite intervals. Since the sum of all $P_{ij}^{(n)}(t)$ is a probability and thus bounded, we conclude by majorized convergence that also $P(t)$ is Lebesgue integrable with respect to t over finite intervals.

Now we can state the recursion

$$P_{ij}(t) = e^{-\lambda_i \cdot t} \cdot \delta_{ij} + \int_0^t e^{-\lambda_i \cdot s} \lambda_i \sum_{k \in E} p_{ik} P_{kj}(t - s)\, ds$$

which results from conditioning on the time s of the first jump from state i. We obtain further

$$P_{ij}(t) = e^{-\lambda_i \cdot t} \cdot \left(\delta_{ij} + \int_0^t e^{+\lambda_i \cdot u} \lambda_i \sum_{k \in E} p_{ik} P_{kj}(u)\, du \right)$$

by substituting $u = t - s$ in the integral. Since $\sum_{k \in E} p_{ik} = 1$ is bounded, we conclude that $P(t)$ is continuous in t. Further, we can differentiate $P(t)$ as given in the recursion and obtain

$$\frac{dP_{ij}(t)}{dt} = -\lambda_i e^{-\lambda_i \cdot t} \cdot \left(\delta_{ij} + \int_0^t f(u) \, du \right) + e^{-\lambda_i \cdot t} \cdot f(t)$$

with f denoting the integrand function. This means nothing else than

$$\frac{dP_{ij}(t)}{dt} = -\lambda_i P_{ij}(t) + \lambda_i \sum_{k \in E} p_{ik} P_{kj}(t)$$

$$= -\lambda_i (1 - p_{ii}) \cdot P_{ij}(t) + \sum_{k \neq i} g_{ik} P_{kj}(t)$$

and thus proves the backward equations. For the forward equations, one only needs to use the Chapman–Kolmogorov equations and apply the backward equations in

$$\frac{dP_{ij}(t)}{dt} = \lim_{h \to 0} \frac{P_{ij}(t+h) - P_{ij}(t)}{h} = \lim_{h \to 0} \sum_{k \in E} P_{ik}(t) \frac{P_{kj}(h) - \delta_{kj}}{h}$$

$$= \sum_{k \in E} P_{ik}(t) \lim_{h \to 0} \frac{P_{kj}(h) - P_{kj}(0)}{h} = \sum_{k \in E} P_{ik}(t) g_{kj}$$

which holds for all $i, j \in E$.
$\square$

Theorem 3.7 *The transition probability matrices can be expressed in terms of the generator by*

$$P(t) = e^{G \cdot t} := \sum_{n=0}^{\infty} \frac{t^n}{n!} G^n$$

for all $t \geq 0$, with G^n denoting the nth power of the matrix G.

Proof: First we validate the solution by

$$\frac{d}{dt} e^{G \cdot t} = \frac{d}{dt} \sum_{n=0}^{\infty} \frac{t^n}{n!} G^n = \sum_{n=1}^{\infty} G^n \frac{d}{dt} \frac{t^n}{n!} = \sum_{n=1}^{\infty} G^n \frac{t^{n-1}}{(n-1)!} = G e^{G \cdot t}$$

which holds for all $t \geq 0$. Furthermore, it is obvious that

$$G e^{G \cdot t} = G \sum_{n=0}^{\infty} \frac{t^n}{n!} G^n = \left(\sum_{n=0}^{\infty} \frac{t^n}{n!} G^n \right) G = e^{G \cdot t} G$$

and thus $P(t) = e^{G \cdot t}$ is a solution of Kolmogorov's forward and backward equations.

Now we show uniqueness of the solution. Let $\tilde{P}(t)$ denote another solution of the forward equations. The differential equations with initial condition translate into the integral equations

$$P(t) = I_E + \int_0^t P(u)G \, du \qquad \text{and} \qquad \tilde{P}(t) = I_E + \int_0^t \tilde{P}(u)G \, du$$

Define a norm for matrices $M = (m_{ij})_{i,j \in E}$ on E by

$$\|M\| := \sup \left\{ \sum_{j \in E} |m_{ij}| : i \in E \right\}$$

Then $\|G\| \leq 2 \cdot \hat{\lambda}$ and $\|AB\| \leq \|A\| \cdot \|B\|$ for any two matrices A and B on E. Further we obtain

$$\left\| P(t) - \tilde{P}(t) \right\| = \left\| \int_0^t P(u) - \tilde{P}(u) \, du \, G \right\|$$

$$\leq \int_0^t \left\| P(u) - \tilde{P}(u) \right\| \, du \cdot \|G\| \qquad (3.2)$$

$$\leq \Delta_t \cdot t \cdot \|G\| \qquad (3.3)$$

with $\Delta_t := \sup\{\|P(u) - \tilde{P}(u)\| : u \leq t\}$, which is finite, since for all $u \geq 0$ we know that $\|P(u)\| = \|\tilde{P}(u)\| = 1$. Plugging the result (3.3) into the right hand of the bound (3.2) again (with time u instead of t), we obtain

$$\left\| P(t) - \tilde{P}(t) \right\| \leq \int_0^t \Delta_t \cdot u \cdot \|G\| \, du \cdot \|G\| = \Delta_t \cdot \frac{t^2}{2} \cdot \|G\|^2$$

Likewise, n–fold repetition of this step achieves the bound

$$\left\| P(t) - \tilde{P}(t) \right\| \leq \Delta_t \cdot \frac{t^n}{n!} \cdot \|G\|^n \leq \Delta_t \cdot \frac{(2\hat{\lambda} \cdot t)^n}{n!}$$

which in the limit $n \to \infty$ yields $0 \leq \left\| P(t) - \tilde{P}(t) \right\| \leq 0$ and consequently $P(t) = \tilde{P}(t)$. As t has been chosen arbitrarily, the statement is proven.
$\square$

Hence the generator of a Markov process uniquely determines all its transition matrices. This can also be seen from the definition, if we agree (without loss of generality) upon the convention $p_{ii} = 0$ for all $\in E$. Then the parameters for the definition of the Markov process can be recovered by

$$\lambda_i = -g_{ii} \quad \text{and} \quad p_{ij} = \frac{g_{ij}}{-g_{ii}}$$

for all $i \neq j \in E$.

However, as in the discrete time case of Markov chains, Markov processes are not completely determined by their transition probability matrices only. The missing link to a complete characterization again is given by the **initial distribution** π with $\pi_i = \mathbb{P}(Y_0 = X_0 = i)$ for all $i \in E$. Then we can express all **finite–dimensional marginal distributions** as in

Theorem 3.8 *For a Markov process $\mathcal{Y}$ with initial distribution π and time instances $0 < t_1 < \ldots < t_n$, $n \in \mathbb{N}$, the equation*

$$\mathbb{P}(Y_{t_1} = j_1, \ldots, Y_{t_n} = j_n)$$
$$= \sum_{i \in E} \pi_i P_{i,j_1}(t_1) P_{j_1,j_2}(t_2 - t_1) \ldots P_{j_{n-1},j_n}(t_n - t_{n-1})$$

holds for all $j_1, \ldots, j_n \in E$.

The proof is left as an exercise. Thus a Markov process $\mathcal{Y}$ with transition probability matrices $(P(t) : t \geq 0)$ admits a variety of versions depending on the initial distribution π. Any such **version** shall be denoted by $\mathcal{Y}^\pi$.

2.   Stationary Distribution

From now on we shall convene on the technical assumption

$$\check{\lambda} := \inf\{\lambda_i : i \in E\} > 0$$

which holds for all queueing systems that we will examine. Then a Markov process is called **irreducible, transient, recurrent** or **positive recurrent** if the defining Markov chain is.

An initial distribution π is called **stationary** if the process $\mathcal{Y}^\pi$ is stationary, i.e. if

$$\mathbb{P}(Y_{t_1}^\pi = j_1, \ldots, Y_{t_n}^\pi = j_n) = \mathbb{P}(Y_{t_1+s}^\pi = j_1, \ldots, Y_{t_n+s}^\pi = j_n)$$

for all $n \in \mathbb{N}$, $0 \leq t_1 < \ldots < t_n$, and states $j_1, \ldots, j_n \in E$, and $s \geq 0$.

Theorem 3.9 *A distribution π on E is stationary if and only if $\pi G = 0$ holds.*

Proof: First we obtain

$$\pi P(t) = \pi e^{G \cdot t} = \sum_{n=0}^{\infty} \frac{t^n}{n!} \pi G^n = \pi I_E + \sum_{n=1}^{\infty} \frac{t^n}{n!} \pi G^n = \pi + \mathbf{0} = \pi$$

for all $t \geq 0$, with $\mathbf{0}$ denoting the zero measure on E. With this, theorem 3.8 yields

$$
\begin{aligned}
\mathbb{P}(Y_{t_1}^{\pi} &= j_1, \dots, Y_{t_n}^{\pi} = j_n) \\
&= \sum_{i \in E} \pi_i P_{i,j_1}(t_1) P_{j_1,j_2}(t_2 - t_1) \dots P_{j_{n-1},j_n}(t_n - t_{n-1}) \\
&= \pi_{j_1} P_{j_1,j_2}(t_2 - t_1) \dots P_{j_{n-1},j_n}(t_n - t_{n-1}) \\
&= \sum_{i \in E} \pi_i P_{i,j_1}(t_1 + s) P_{j_1,j_2}(t_2 - t_1) \dots P_{j_{n-1},j_n}(t_n - t_{n-1}) \\
&= \mathbb{P}(Y_{t_1+s}^{\pi} = j_1, \dots, Y_{t_n+s}^{\pi} = j_n)
\end{aligned}
$$

for all times $t_1 < \dots < t_n$ with $n \in \mathbb{N}$, and states $j_1, \dots, j_n \in E$. Hence the process $\mathcal{Y}^{\pi}$ is stationary.

On the other hand, if π is a stationary distribution, then we necessarily obtain $\pi P(t) = \pi e^{G \cdot t} = \pi$ for all $t \geq 0$. As above, this means $\sum_{n=1}^{\infty} \frac{t^n}{n!} \pi G^n = \mathbf{0}$ for all $t \geq 0$, which yields $\pi G = 0$ because of the uniqueness of the zero power series.

$\square$

By definition of the generator G and equation (3.1), the equation $\pi G = 0$ is equivalent to an equation system

$$\sum_{i \neq j} \pi_i g_{ij} = -\pi_j g_{jj} \qquad \Longleftrightarrow \qquad \sum_{i \neq j} \pi_i g_{ij} = \pi_j \sum_{i \neq j} g_{ji} \qquad (3.4)$$

for all $j \in E$. This system can be intepreted as follows. We call the value $\pi_i g_{ij}$ **stochastic flow** from state i to state j in equilibrium. Then the above equations mean that the accrued stochastic flow into any state j equals the flow out of this state. Equations (3.4) are called the (global) **balance equations**.

Example 3.10 The generator of the Poisson process with parameter λ (see example 3.1) is given by

$$
G = \begin{pmatrix}
-\lambda & \lambda & 0 & 0 & \dots \\
0 & -\lambda & \lambda & 0 & \ddots \\
0 & 0 & -\lambda & \lambda & \ddots \\
\vdots & \ddots & \ddots & \ddots & \ddots
\end{pmatrix}
$$

This process has no stationary distribution, which can be seen as follows. The balance equations for the Poisson process are given by

$$\pi_0 \lambda = 0 \qquad \text{and} \qquad \pi_i \lambda = \pi_{i-1} \lambda$$

for all $i \geq 1$. It is immediately evident that these are solvable only by $\pi_i = 0$ for all $i \in E$, which means that there is no stationary distribution π.

The question of existence and uniqueness of a stationary distribution for $\mathcal{Y}$ can be reduced to the same question for $\mathcal{X}$, which we have examined in the preceding chapter:

Theorem 3.11 *Let the underlying Markov chain $\mathcal{X}$ in the definition of the Markov process $\mathcal{Y}$ be irreducible and positive recurrent. Further assume that $\check{\lambda} := \inf\{\lambda_i : i \in E\} > 0$. Then there is a unique stationary distribution for $\mathcal{Y}$.*

Proof: According to theorems 2.25 and 2.18, the transition matrix P of $\mathcal{X}$ admits a unique stationary distribution ν with $\nu P = \nu$. The generator G is defined by $G = \Lambda(P - I_E)$, with $\Lambda = diag(\lambda_i : i \in E)$. Hence the measure $\mu := \nu \Lambda^{-1}$ is stationary for $\mathcal{Y}$. Since $\check{\lambda} > 0$, the measure μ is finite, with total mass bounded by $\check{\lambda}^{-1} < \infty$. Now the normalization

$$\pi_j := \frac{\mu_j}{\sum_{i \in E} \mu_i} = \frac{\nu_j / \lambda_j}{\sum_{i \in E} \nu_i / \lambda_i} \tag{3.5}$$

for all $j \in E$ yields a stationary distribution for $\mathcal{Y}$. This is unique because ν is unique and the construction of π from ν is reversible.
$\square$

Finally we give two important results for the asymptotic behaviour of a Markov process. These shall be proven in chapter 7 (see example 7.13). We call a Markov process **regular** if it satisfies the conditions given in the preceding theorem. If $\mathcal{Y}$ is a regular Markov process, then the limit

$$\lim_{t \to \infty} \mathbb{P}(Y_t = j) = \pi_j \tag{3.6}$$

of the marginal distribution at time t tends to the stationary distribution as t tends to infinity. Further the limit

$$\lim_{t \to \infty} P_{ij}(t) = \pi_j \tag{3.7}$$

holds for all $i, j \in E$ and is independent of i.

Notes

An early text book on Markov processes with discrete state space is Chung [27]. Other classical text book presentation are Karlin and Taylor [46], Breiman [16], or Çinlar [25]. An exposition on non–homogeneous Markov processes on discrete state spaces can be found under the name Markov jump processes in Gikhman and Skorokhod [39, 38].

Exercise 3.1 Consider a population of male and female species. There is an infinitesimal rate $\lambda > 0$ that any male and female produce a single offspring, which will be female with probability p. Determine a Markov process which models the numbers F_t and M_t of female and male species at any time t.

Exercise 3.2 Let X and Y denote two independent random variables which are distributed exponentially with parameters λ and μ, respectively. Prove the following properties:
(a) $X \neq Y$ almost certainly.
(b) The random variable $Z := \min\{X, Y\}$ is distributed exponentially with parameter $\lambda + \mu$.
(c) $\mathbb{P}(X < Y) = \lambda/(\lambda + \mu)$

Exercise 3.3 Let $\mathcal{Y}^{(1)}$ and $\mathcal{Y}^{(2)}$ denote independent Poisson processes with intensities λ_1 and λ_2, respectively. Show that the process $\mathcal{Y} = (Y_t : t \in \mathbb{R}_0^+)$ defined by $Y_t = Y_t^{(1)} + Y_t^{(2)}$ for all $t \geq 0$ is a Poisson process with intensity $\lambda = \lambda_1 + \lambda_2$. The process $\mathcal{Y}$ is called the **superposition** of $\mathcal{Y}^{(1)}$ and $\mathcal{Y}^{(2)}$.

Exercise 3.4 Prove theorem 3.8.

Exercise 3.5 Determine the finite–dimensional marginal distributions for a Poisson process with parameter λ.

Exercise 3.6 Let $\mathcal{Y}$ denote a Poisson process with parameter λ. Given that there is exactly one arrival in the interval $[0, t]$, show that the exact time of the arrival within $[0, t]$ is uniformly distributed.

Exercise 3.7 Verify the Chapman–Kolmogorov equations for a Poisson process.

Chapter 4

MARKOVIAN QUEUES IN CONTINUOUS TIME

The methods of analyzing Markov processes are already sufficient for the treatment of quite a variety of queueing systems. These are commonly known as elementary or **Markovian queues**. The most classical of them shall be examined in this chapter.

1. The M/M/1 Queue

The M/M/1 queue in continuous time is defined by the following characteristics: The arrival process is a Poisson process with some rate $\lambda > 0$. The service times are iid and distributed exponentially with service rate $\mu > 0$. There is one server and the service discipline is first come first served (FCFS, see example 1.1).

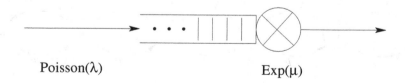

Poisson(λ)   Exp(μ)

Figure 4.1. M/M/1 queue

For the Poisson process, the inter–arrival times are distributed exponentially with parameter λ. Since the exponential distribution is memoryless, the system process $\mathcal{Q} = (Q_t : t \in \mathbb{R}_0^+)$ can be modelled by a Markov process with state space $E = \mathbb{N}_0$ and generator

$$G = \begin{pmatrix} -\lambda & \lambda & 0 & 0 & \dots \\ \mu & -\lambda-\mu & \lambda & 0 & \ddots \\ 0 & \mu & -\lambda-\mu & \lambda & \ddots \\ \vdots & \ddots & \ddots & \ddots & \ddots \end{pmatrix}$$

Here, the first line represents the possible transitions if the system is empty. In this case there can only occur single arrivals according to the Poisson process with rate λ. If the system is not empty, there are two possibilities: Either an arrival occurs (with rate λ) or a service is completed (with rate μ). Contrary to the M/M/1 queue in discrete time, arrivals and service completions cannot occur at the same time. This follows from the memoryless property of the exponential distribution and exercise 3.2. The parameter of the holding time for the states of a non–empty system is explained by exercise 3.2.

Clearly, the structure of the matrix G shows that the process $\mathcal{Q}$ is irreducible and hence there is at most one stationary distribution π for $\mathcal{Q}$. According to theorem 3.9, this must satisfy $\pi G = 0$, which translates into the system

$$\pi_0 \lambda = \pi_1 \mu \tag{4.1}$$
$$\pi_n(\lambda + \mu) = \pi_{n-1}\lambda + \pi_{n+1}\mu \qquad \text{for all} \quad n \geq 1 \tag{4.2}$$
$$\sum_{n=0}^{\infty} \pi_n = 1 \tag{4.3}$$

of equations, where the latter is simply the normalization of the distribution π. The first two equations are the global **balance equations** and can be illustrated by the following scheme:

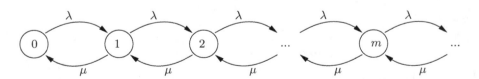

Figure 4.2.   Transition rates for the M/M/1 queue

This gives the rates of jumps between the states of the system. If we encircle any one state, then the sum of the rates belonging to the arcs reaching into this state must equal the sum of the rates which belong to the arcs that go out of this state. If this is the case, then we say that the system is in balance. The conditions for this are given in equations (4.1) and (4.2).

The solution of the above system of equations can be obtained by the following considerations: The first equation yields

$$\pi_1 = \pi_0 \frac{\lambda}{\mu} =: \pi_0 \rho$$

with $\rho := \lambda/\mu$. By induction on n we obtain from the second equation

$$\pi_{n+1} = \frac{1}{\mu}(\pi_n(\lambda + \mu) - \pi_{n-1}\lambda) = \pi_n \frac{\lambda}{\mu} + \pi_n - \pi_{n-1}\frac{\lambda}{\mu}$$

$$= \pi_n \rho$$

for all $n \in \mathbb{N}$, where the last equality holds by induction hypothesis. Thus the geometric approach $\pi_n = \pi_0 \rho^n$ for all $n \in \mathbb{N}_0$ solves the first two equations. The last equation now yields

$$1 = \sum_{n=0}^{\infty} \pi_n = \pi_0 \sum_{n=0}^{\infty} \rho^n = \frac{1}{1-\rho}\pi_0$$

if and only if $\rho < 1$, which means $\lambda < \mu$. Hence there is a stationary distribution of the system, given by

$$\pi_n = (1 - \rho)\rho^n$$

for all $n \in \mathbb{N}_0$, if and only if the so–called queue **load** $\rho = \lambda/\mu$ remains smaller than one.

In this case several performance measures of the queueing system can be derived immediately. All of them are computed by means of the stationary distribution. Thus they hold only for the system being in equilibrium, which is attained asymptotically.

For instance, the probability that the system is empty is given by $\pi_0 = 1 - \rho$. The mean and the variance of the number N of users in the system are given as

$$\mathbb{E}(N) = \sum_{n=1}^{\infty} n\pi_n = (1 - \rho) \sum_{n=1}^{\infty} n\rho^n = \frac{\rho}{1-\rho}$$

and $\mathbb{V}ar(N) = \rho/(1-\rho)^2$. The probability R_K that there are at least K users in the system is

$$R_K = \sum_{n=K}^{\infty} \pi_n = (1 - \rho) \sum_{n=K}^{\infty} \rho^n = \rho^K$$

As expected, these equations show that with increasing load $\rho \to 1$ the mean number of users in the system grows and the probability of an idle system decreases.

2. Skip–Free Markov Processes

There are many variations of the M/M/1 queue which can be analyzed by the same method. In order to show this we first put the analysis presented in the preceding section in a more general context. This will be applicable to a large variety of queueing models.

The Markov process which models the M/M/1 queue has the decisive property that transitions are allowed to neighbouring states only, i.e. $g_{ij} = 0$ for states $i, j \in \mathbb{N}_0$ with $|i - j| > 1$. The result is a very simple state transition graph of a linear form and correspondingly a set of balance equations, given by (4.1) and (4.2), which can be solved easily. We can retain the same method of analysis if we relax the special assumption that $g_{i,i+1}$ and $g_{i,i-1}$ be independent of i.

Thus we define a **skip–free Markov process** by the property that its generator $G = (g_{ij})_{i,j\in E}$ satisfies $g_{ij} = 0$ for all states $i, j \in E \subset \mathbb{N}_0$ with $|i - j| > 1$. For queueing systems this means that there are only single arrivals or departures. Thus every Markovian queueing system with single arrivals and departures can be modelled by a skip–free Markov process.

Denote the remaining infinitesimal transition rates by

$$\lambda_i := g_{i,i+1} \qquad \text{and} \qquad \mu_i := g_{i,i-1}$$

for all possible values of i. The rates λ_i and μ_i are called **arrival rate**s and **departure rate**s, respectively. The state transition graph of such a process assumes the form

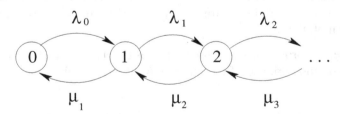

Figure 4.3.   A skip–free Markov process

Its balance equations are given by $\lambda_0 \pi_0 = \mu_1 \pi_1$ and

$$(\lambda_i + \mu_i)\pi_i = \lambda_{i-1}\pi_{i-1} + \mu_{i+1}\pi_{i+1}$$

for all $i \in \mathbb{N}$. By induction on i it is easily shown that these are equivalent to the equation system

$$\lambda_{i-1}\pi_{i-1} = \mu_i \pi_i \tag{4.4}$$

for all $i \in \mathbb{N}$. This system is solved by successive elimination with a solution of the form

$$\pi_i = \pi_0 \prod_{j=0}^{i-1} \frac{\lambda_j}{\mu_{j+1}} = \pi_0 \frac{\lambda_0 \lambda_1 \cdots \lambda_{i-1}}{\mu_1 \mu_2 \cdots \mu_i} \tag{4.5}$$

for all $i \geq 1$. The solution π is a probability distribution if and only if it can be normalized, i.e. if $\sum_{n \in E} \pi_n = 1$. This condition implies

$$1 = \sum_{n \in E} \pi_0 \prod_{j=0}^{n-1} \frac{\lambda_j}{\mu_{j+1}} = \pi_0 \sum_{n \in E} \prod_{j=0}^{n-1} \frac{\lambda_j}{\mu_{j+1}}$$

with the empty product being defined as one. This means that

$$\pi_0 = \left(\sum_{n \in E} \prod_{j=0}^{n-1} \frac{\lambda_j}{\mu_{j+1}} \right)^{-1} \tag{4.6}$$

and thus π is a probability distribution if and only if the series in the brackets converges. In this case, the stationary distribution of a skip–free Markov process is given by (4.6) and (4.5).

3. The M/M/∞ Queue

The first application of the analysis of the last section to a queueing system shall be the M/M/∞ queue. This is a queue without queueing: There are infinitely many servers such that every incoming user finds an idle server immediately. Arrivals are governed by a Poisson process with intensity $\lambda > 0$, and the service times are exponentially distributed with rate $\mu > 0$, equal for each server. Due to lemma 3.2, the system process is Markovian. Furthermore, there are only single arrivals and departures. Hence the M/M/∞ queue can be modelled by a skip–free Markov process.

Since the arrival process is independent of the rest of the queue, the arrival rates of the respective skip–free Markov process are constant. In the notation of section 2 we can thus specify $\lambda_n = \lambda$ for all $n \in \mathbb{N}_0$. Departures occur upon service completions. According to lemma 3.2 and due to the memoryless property of the exponential distribution (see lemma 3.2), the departure rates are given by $\mu_n = n \cdot \mu$ for all $n \in \mathbb{N}$.

Define $\rho := \lambda / \mu$. Then the series in (4.6) assumes the value

$$\sum_{n=0}^{\infty} \prod_{j=0}^{n-1} \frac{\lambda_j}{\mu_{j+1}} = \sum_{n=0}^{\infty} \frac{\rho^n}{n!} = e^{\rho}$$

and thus converges regardless of the value of ρ. This means that the M/M/∞ queue always has a stationary distribution, which is not surprising as infinitely many servers cannot be exhausted, whatever the arrival intensity amounts to.

Due to formulae (4.6) and (4.5), we obtain the stationary distribution π as given by $\pi_0 = e^{-\rho}$ and

$$\pi_n = e^{-\rho} \cdot \frac{\rho^n}{n!}$$

for all $n \in \mathbb{N}$, which is a Poisson distribution with parameter ρ. Hence the mean and the variance of the number N of users in the stationary system are given by $\mathbb{E}(N) = \mathbb{V}ar(N) = \rho$.

Since there is no queueing in the M/M/∞ system, all waiting times are zero and the mean sojourn time in the system equals $1/\mu$. This means that all users passing through such a system are independently kept there for an exponentially distributed time. In the context of queueing networks (see chapter 5), the M/M/∞ queue is therefore often called an (independent) **delay system**.

4.   The M/M/k Queue

The M/M/k queue is provided with k identical servers which can serve users in parallel. Users arrive according to a Poisson process with intensity $\lambda > 0$, and the service time distribution is exponential with parameter $\mu > 0$ at all servers. Whenever a user arrives and finds all servers busy (i.e. at least k users in the system) he queues up in the waiting room. From there the next waiting user is served in the order of a FIFO discipline as soon as one of the servers becomes idle. An arriving user finding less than k users already in the system (i.e. there are idle servers at the time of arrival) chooses any server and starts service immediately.

For this type of queue the dynamics is a mixture between the M/M/∞ queue and the M/M/1 queue. Up to the value of k users in the system, the service (and thus the departure) rate increases like $\mu_n = n \cdot \mu$ for $1 \leq n \leq k$. Starting from k users in the system there are no servers anymore to keep up with newly arriving users, and the departure rate remains $\mu_n = k \cdot \mu$ for all $n \geq k+1$. The independence of the arrival process yields constant arrival rates $\lambda_n = \lambda$ for all $n \in \mathbb{N}_0$.

Again we define $\rho := \lambda/\mu$. The series in (4.6) specifies to

$$\sum_{n=0}^{\infty} \prod_{j=0}^{n-1} \frac{\lambda_j}{\mu_{j+1}} = \sum_{n=0}^{k-1} \frac{\rho^n}{n!} + \frac{\rho^k}{k!} \sum_{n=0}^{\infty} \left(\frac{\rho}{k}\right)^n$$

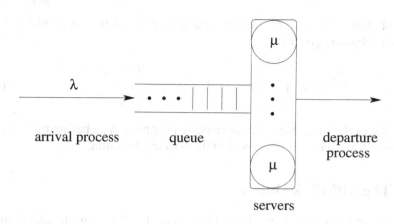

Figure 4.4. M/M/k queue

which is finite if and only if $\rho < k$. In this case the stationary distribution π is given by formulae (4.6) and (4.5) as

$$\pi_0 = \left(\sum_{n=0}^{k-1} \frac{\rho^n}{n!} + \frac{\rho^k}{(k-1)! \cdot (k-\rho)} \right)^{-1}$$

and

$$\pi_n = \pi_0 \cdot \frac{\rho^n}{n!}, \qquad 1 \leq n \leq k$$

$$\pi_n = \pi_k \cdot \left(\frac{\rho}{k} \right)^{n-k}, \qquad n > k$$

Here we see the M/M/∞ form for $n \leq k$ and the M/M/1 form beginning with $n \geq k$, where π_k substitutes the base value that is played by π_0 for the pure M/M/1 queue.

The fact that the M/M/k queue behaves for more than k users in the system like an M/M/1 queue with load ρ/k is further illustrated by the following observation. Let N denote the number of users in the system that is in equilibrium. Consider the conditional probability $p_n := \mathbb{P}(N = n | N \geq k)$ for $n \geq k$. This is computed as

$$p_n = \frac{\pi_n}{\sum_{i=k}^{\infty} \pi_i} = \pi_k \left(\frac{\rho}{k} \right)^{n-k} \bigg/ \pi_k \sum_{i=k}^{\infty} \left(\frac{\rho}{k} \right)^{i-k} = \left(\frac{\rho}{k} \right)^{n-k} \left(1 - \frac{\rho}{k} \right)$$

Since $n - k$ is the number N_q of users waiting in the queue, the conditional distribution of N_q given that all servers are busy has exactly the same (geometric) form as the stationary distribution for the M/M/1 system process.

The probability $\mathbb{P}\{N \geq k\}$ of the conditioning event that in equilibrium all servers are busy is given by

$$\sum_{n=k}^{\infty} \pi_n = \left(1 + (k-1)! \cdot (k-\rho) \cdot \sum_{n=0}^{k-1} \frac{\rho^{n-k}}{n!}\right)^{-1} \qquad (4.7)$$

This is the probability that a newly arriving user must wait before he is served. The above formula for it is called **Erlang's delay formula**.

5. The M/M/k/k Queue

In stochastic modelling there always is a trade–off between the adaptation of the model to reality and its simplicity, i.e. its analytic tractability. We have seen that the nicest solutions could be derived for the M/M/1 queue (a geometric distribution) and the M/M/∞ queue (a Poisson distribution). The solution for the M/M/k queue, which is more realistic for most practical applications, is also more involved. For all these models we kept the often unrealistic assumption of an infinite waiting room. The models in this and the following sections stem from more realistic specifications. Historically, they belong to the first applications which founded the field of queueing theory.

In the times of A.K. Erlang, at the beginning of the 20th century, telephone calls had to be connected by an operator. The telephone companies installed call centers where a number k of operators served call requests which arrived from a large number of subscribers. Whenever all operators are busy with serving call requests and a new subscriber calls to get a line, this subscriber will be rejected.

If we model the arriving requests by a Poisson process and the duration of the operators' services by an exponential distribution, then we get an M/M/k/k queue as a model for this application. The subscribers with their call requests are the users and the operators are the servers. There are k servers and as many places in the system, i.e. there is no additional waiting room.

Let the intensity of the Poisson arrival process and the rate of the exponential service times be denoted by $\lambda > 0$ and $\mu > 0$, respectively. Again we can use a skip–free Markov process to analyze this system. In this notation, we obtain $\lambda_n = \lambda$ for all $n = 0, \ldots, k-1$ and $\mu_n = n \cdot \mu$ for $n = 1, \ldots, k$. The values of λ_n and μ_n are zero for all other indices n. Define $\rho := \lambda/\mu$. The series in (4.6) is in this case

$$\sum_{n \in E} \prod_{j=0}^{n-1} \frac{\lambda_j}{\mu_{j+1}} = \sum_{n=0}^{k} \frac{\rho^n}{n!}$$

which is finite, regardless of the value for ρ. Hence a stationary distribution π always exists and is given by

$$\pi_0 = \left(\sum_{n=0}^{k} \frac{\rho^n}{n!}\right)^{-1} \quad \text{and} \quad \pi_n = \pi_0 \cdot \frac{\rho^n}{n!}$$

for all $n = 1, \ldots, k$. The main performance measure for this application is the probability that all operators are busy and the company is unable to accept new call requests. This is given by

$$\pi_k = \frac{\rho^k}{k!} \left(\sum_{n=0}^{k} \frac{\rho^n}{n!}\right)^{-1}$$

which of course is valid only under the stationary regime, i.e. in equilibrium. This expression is known as **Erlang's loss formula**.

Note that the expression of π_0 for the M/M/∞ queue is the limit of the respective expression for the M/M/k/k model as k tends to infinity. Even further, the stationary distribution for the M/M/k/k queue converges to the stationary distribution of the M/M/∞ for increasing k.

6.   The M/M/k/k+c/N Queue

A simplifying assumption in the previous model has been the constant arrival rates $\lambda_n = \lambda$. This implies that even for a high number of users in the queue the intensity of new arrivals does not diminish. While this is a reasonable assumption for an application to call centers, where the number of operators (and thus the maximal number of users in the system) is only marginal compared to the number of all subscribers, there are other applications for which such an assumption would not be realistic.

Consider a closed computer network with k servers and N terminals. Every terminal sends a job to one of the servers after some exponentially distributed think time. If a server is available, i.e. idle, then this job is served, demanding an exponential service time. A terminal that has a job in a server may not send another job request during the service time. Whenever a terminal sends a job request and all servers are busy at that time, then the job is put into a queue. This queue has maximal capacity c, i.e. if a terminal sends a job request and the queue is already filled with c jobs, then this new job request is rejected and the terminal starts another think time.

This application can be modelled by an M/M/k/k+c/N queue if we interpret the users in the system as the job requests that are in service or waiting. Denote

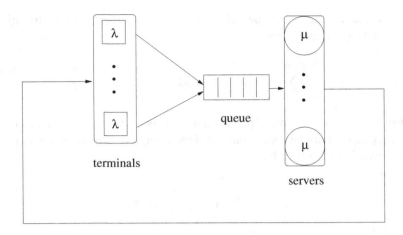

Figure 4.5.   A closed computer network

the parameters of the exponential think time and service time distributions by $\lambda > 0$ and $\mu > 0$, respectively. Without loss of generality we may assume that $k + c \leq N$. Then the queue in consideration is a skip–free Markov process with arrival rates $\lambda_n = (N - n) \cdot \lambda$ for $n = 0, \ldots, k + c - 1$ and departure rates $\mu_n = \min(n, k) \cdot \mu$ for $n = 1, \ldots, k + c$. As always, define $\rho := \lambda/\mu$. The series in (4.6) amounts to

$$\sum_{n \in E} \prod_{j=0}^{n-1} \frac{\lambda_j}{\mu_{j+1}} = \sum_{n=0}^{k} \binom{N}{n} \cdot \rho^n + \sum_{n=k+1}^{k+c} \frac{N! \cdot \rho^n}{(N-n)! \cdot k! \cdot k^{n-k}} \tag{4.8}$$

and thus is finite for every value of ρ. The stationary distribution π is given by

$$\pi_0 = \left(\sum_{n=0}^{k} \binom{N}{n} \cdot \rho^n + \sum_{n=k+1}^{k+c} \frac{N! \cdot \rho^n}{(N-n)! \cdot k! \cdot k^{n-k}} \right)^{-1}$$

and

$$\pi_n = \pi_0 \cdot \binom{N}{n} \cdot \rho^n, \qquad 1 \leq n \leq k$$

$$\pi_n = \pi_0 \cdot \frac{N! \cdot \rho^n}{(N-n)! \cdot k! \cdot k^{n-k}}, \qquad k + 1 \leq n \leq k + c$$

There are several interesting special cases. For $c = 0$ there is no room for a queue of waiting jobs. Then the stationary distribution simplifies to a binomial distribution with parameters (N, p), where $p = \rho/(1 + \rho)$, which is truncated to the states $n = 0, \dots, k$. Such a distribution is called an **Engset distribution**.

For $c = N - k$ the queue has an important application in reliability theory. This is known as the **machine repair problem**. In a production site there are N machines which are prone to failure. Each of them breaks down after a working time which is exponentially distributed with parameter λ. There are k repairmen that take care of the broken machines sequentially. The repair times are exponential with parameter μ. Then the system process of the $M/M/k/N/N$ queue yields the number of broken machines.

Notes

The models presented in this chapter are the oldest within queueing theory. Applications to telephone networks date back to the beginning of the 20th century, notably Erlang [33] and Engset [32].

Skip–free Markov processes have been extensively used for populations models. Therefore the name birth–and–death processes is very popular for them, with λ_i and μ_i denoting the transition rates for a birth and a death, respectively, if the population has i members. However, the authors think that such a name is inappropriate for queueing models and thus prefer the more technical term skip–free.

For more Markovian queueing models see Kleinrock [50]. An analysis of non–homogeneous (namely periodic) Markovian queues is given in Breuer [17, 22].

Exercise 4.1 Verify the formula $\mathbb{V}ar(N) = \rho/(1 - \rho)^2$ for the stationary variance of the number of users in the M/M/1 queue.

Exercise 4.2 Show that the equation system (4.4) is equivalent to the balance equations for a skip–free Markov process. Prove the form (4.5) of its solution.

Exercise 4.3 Prove Erlang's delay formula (4.7).

Exercise 4.4 Compare the stationary mean number of users in the system for the following three queueing systems: (a) an M/M/1 queue with arrival intensity λ and service rate μ, (b) an M/M/2 system with arrival intensity λ and service rate $\mu/2$, and (c) two independent M/M/1 queues with arrival intensity $\lambda/2$ to each of them and equal service rate μ. Explain the differences.

Exercise 4.5 Explain equation (4.8).

Exercise 4.6 Show that the stationary distribution for an $M/M/k/k/N$ queue is an Engset distribution.

Exercise 4.7 Analyze the M/M/1/c queue with arrival intensity λ and service rate μ. This always has a stationary distribution π. Show that in the limit $c \rightarrow \infty$, there are two possibilities: Either $\rho < 1$ and π converges to the stationary distribution of the M/M/1 queue, or $\rho \geq 1$ and π converges to the zero measure.

Exercise 4.8 Examine the M/M/1 queue with users who are discouraged by long queue lengths. This can be modelled by arrival rates $\lambda_n = \lambda/(n+1)$ for all $n \in \mathbb{N}_0$. Show that the stationary distribution is Poisson.

Chapter 5

MARKOVIAN QUEUEING NETWORKS

A set of interconnected queueing stations in which any user, upon departing from one station, can join another or must leave the total system is called a **queueing network**. The paths along which a user may travel from station to station are determined by **routing probabilities** q_{ij}. Travel times, in general, are assumed to be zero.

A queueing network may be regarded as a directed graph whose nodes represent the stations, and whose edges represent links between nodes. Between nodes i and j an edge exists if and only if the routing probability q_{ij}, i.e. the probability to join station j after service completion at station i, is greater than zero. There may be also links from and to the outside of the network, representing the possibility for users to enter or leave the system. Let q_{j0} denote the probability for a user to depart from the network after being served at node j. Then $\sum_{k=0}^{M} q_{jk} = 1$, with M the number of stations in the network. The matrix $Q = (q_{ij})_{i,j \leq M}$ is called the **routing matrix** of the network. Given Q, the probabilities for network departures are implicitly determined by $q_{j0} = 1 - \sum_{k=1}^{M} q_{jk}$.

Routing probabilities may be state dependent, where a network state usually is defined by the vector $\mathbf{n} = (n_1, \ldots, n_M)$ of actual numbers n_i of customers in stations $i = 1, \ldots, M$. More complex state definitions arise when customers of different classes require different amounts of service and follow different routes through the network. It may be the case that a particular routing behaviour is associated with a certain group of classes, while other groups follow different rules. This leads to the notion of **chains** in a network. A chain defines a particular subset (called **category**) of customers who travel through the network according to a particular routing mechanism. Class changes of customers within a chain are possible, but no customer can pass over to some class of another chain. The pair of class and chain identifiers is called a **category index**.

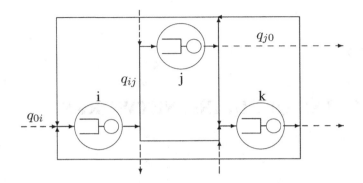

Figure 5.1.   Open Queueing Network

A network is called a **closed network** if there is no traffic entering the network from outside, and all departure probabilities q_{i0} are zero. A network that allows incoming and outgoing traffic is called an **open network**. Since we are only interested in systems that eventually reach equilibrium, the cases with entering but no outgoing traffic, or vice versa, are excluded from investigation. As mentioned above, it is further possible to discriminate between different user classes and different chains in a network. Each chain is associated with its own routing matrix Q_c. In this case the network may be open for some chains, and closed for others. Such a network is called a **mixed network**. In this book we concentrate on the case of state independent routing.

In the simplest case, when customers are non-distinguishable, the dynamic behaviour of a queueing network is best described by a vector-valued stochastic process $(N_t : t \geq 0)$ with state space $\mathbb{N}_0^M$. In case that we consider different customer classes and/or special characteristics of service or even inter-arrival times, a network state, clearly, may be described differently. As a construct for stochastic modelling, queueing networks are subject to performance analysis, the performance measures of interest being similar to those for isolated stations. In most cases one is interested in the average delay (or system time) $\bar{T}$ that a user experiences when travelling through the network, and in the mean throughput $\bar{S}$ as well as the mean total number $\bar{N}$ of customers that are resident. According to Little's result (see theorem 1.9), these quantities are related by

$$\bar{S} \cdot \bar{T} = \bar{N}. \tag{5.1}$$

They can easily be evaluated by means of the stationary state probabilities (that hold in equilibrium), if those exist and are known. For instance, with $p_\mathbf{n}$ denoting the stationary probability for the network to be in state $\mathbf{n} = (n_1, \ldots, n_M)$,

$$\bar{N} = \sum_{\mathbf{n} \in \mathbb{N}_0^M} p_{\mathbf{n}} \cdot \sum_{i=1}^{M} n_i.$$

In general, the calculation of stationary probabilities, if they exist, represents an unsolved or at least intricate problem. This is due to the fact that in most cases no closed form expressions are known. Nevertheless, there are queueing networks for which stationary state probabilities can be obtained by forming the product of all those stationary state probabilities that are associated with the network stations in isolation. Such queueing networks are called **product form (PF-) networks** or **separable networks**. Among separable networks, the best understood are Markovian queueing networks, i.e. networks for which the stochastic process $(N_t : t \geq 0)$ is Markov, and which allow a product form solution. We shall be concerned mainly with the class of PF-networks in the following sections, and shall concentrate on the most simple cases only. For continuing information about queueing networks of more complex structure the reader is referred to the abundant literature on this topic.

1. Balance Equations and Reversibility Properties

Let $\mathcal{N} = (N_t : t \geq 0)$ be a vector-valued continuous time Markov process with state space E that describes a queueing network with M stations. In the simplest case, when there is only one class of customers travelling through the network, and no phase structures need to be considered for service (and inter-arrival) time distributions, the state space E forms a subset of $\mathbb{N}_0^M$.

$\mathcal{N}$ can be considered as a random walk process on the network graph. Let $G = (g_{\mathbf{mn}})_{\mathbf{m},\mathbf{n} \in E}$ denote the generator matrix of $\mathcal{N}$. Then, given that the process is irreducible, it assumes equilibrium if and only if the system of equations

$$\boldsymbol{p} G = \sum_{\mathbf{m} \in E} p_{\mathbf{m}} g_{\mathbf{mn}} = \boldsymbol{o} \tag{5.2}$$

possesses a finite positive solution $\boldsymbol{p} = (p_{\mathbf{n}})_{\mathbf{n} \in E}$ (see theorem 3.9). Any such solution $\boldsymbol{p}$ can be normed as to satisfy $\sum_{\mathbf{n} \in E} p_{\mathbf{n}} = 1$ and to represent the unique stationary distribution of $\mathcal{N}$, i. e. the joint stationary queue length distribution of the network. For indistinguishable customers, irreducibility of $\mathcal{N}$ is equivalent to the possibility for a customer to be able, upon leaving a station i and subsequently travelling through the network, to finally reach any other station j or, in case of an open network, to reach the exterior of the network. Mathematically spoken, this means that there are integers

$k_1 = i, k_2, \ldots, k_n, k_{n+1} = j$ such that $\prod_{\ell=1}^{n} q_{k_\ell, k_{\ell+1}} > 0$, where in case that the exterior is meant by the source or the destination, the respective index i or j has value 0.

Equation (5.2) mirrors a situation that we call **global balance**. A term of the form $p_{\mathbf{m}} g_{\mathbf{mn}}$, where $g_{\mathbf{mn}}$ is the instantaneous transition rate from state $\mathbf{m}$ to state $\mathbf{n}$, is called **probability flux** or **rate of flow** from $\mathbf{m}$ to $\mathbf{n}$. Since G, as a generator matrix, satisfies $\sum_{\mathbf{n} \in E} g_{\mathbf{mn}} = 0$, (5.2) is equivalent to

$$\sum_{\substack{\mathbf{m} \in E \\ \mathbf{m} \neq \mathbf{n}}} p_{\mathbf{m}} g_{\mathbf{mn}} = \sum_{\substack{\mathbf{m} \in E \\ \mathbf{m} \neq \mathbf{n}}} p_{\mathbf{n}} g_{\mathbf{nm}}, \tag{5.3}$$

stating that the probability flux *into* state $\mathbf{n}$ equals the probability flux *out of* state $\mathbf{n}$. As opposed to that, we speak of a **detailed balance equation**, if

$$p_{\mathbf{m}} g_{\mathbf{mn}} = p_{\mathbf{n}} g_{\mathbf{nm}}. \tag{5.4}$$

There are several concepts of balance between rates of flow in Markov processes and, correspondingly, in Markovian queueing networks. These concepts are tightly connected with the property of reversibility in the theory of Markov processes. In order to illustrate this relationship, let us first specify what is meant by a **reversed process** $\mathcal{N}^{(r)} = (N_t^{(r)})_{t \in \mathbb{R}_0^+}$ associated with some Markov process $\mathcal{N} = (N_t)_{t \in \mathbb{R}_0^+}$ with state space E.

The **reversal** $\mathcal{N}^{(r)}$ of the Markov process $\mathcal{N}$ is a process that is developing in time in forward direction just as the original process does in backward direction, on the same state space E, i. e., for some $\tau \in \mathbb{R}_0^+$, we have $N_t^{(r)} = N_{\tau - t} \, \forall \, t \in \mathbb{R}_0^+$. If $\mathcal{N}$ is time-homogeneous and stationary, the value of τ does not matter, and so

$$N_t^{(r)} = N_{-t} \quad \text{for all} \quad t \in \mathbb{R}_0^+.$$

$\mathcal{N}$ is called the **forward process** corresponding to the reversed or **backward process** $\mathcal{N}^{(r)}$. If the forward process $\mathcal{N}$ is time-homogeneous, irreducible, and stationary, then so is the reversed process $\mathcal{N}^{(r)}$.

Let $G = (g_{\mathbf{mn}})_{\mathbf{m}, \mathbf{n} \in E}$ and $G^{(r)} = (g_{\mathbf{mn}}^{(r)})_{\mathbf{m}, \mathbf{n} \in E}$ denote the generator matrices of $\mathcal{N}$ and $\mathcal{N}^{(r)}$, respectively, with total transition rates

$$\gamma_{\mathbf{m}} = \sum_{\substack{\mathbf{n} \in E \\ \mathbf{n} \neq \mathbf{m}}} g_{\mathbf{mn}}, \quad \gamma_{\mathbf{m}}^{(r)} = \sum_{\substack{\mathbf{n} \in E \\ \mathbf{n} \neq \mathbf{m}}} g_{\mathbf{mn}}^{(r)} \quad \text{for any} \ \mathbf{m} \in E.$$

As can easily be seen, the instantaneous transition rates $g_{\mathbf{mn}}$ and $g_{\mathbf{mn}}^{(r)}$ are, in general, not the same for an arbitrary pair of states $\mathbf{m}, \mathbf{n}$. On the other side, given that $\mathbf{p} = (p_\mathbf{n})_{\mathbf{n}\in E}$ and $\mathbf{p}^{(r)} = (p_\mathbf{n}^{(r)})_{\mathbf{n}\in E}$ denote the stationary distribution vectors of $\mathcal{N}$ and $\mathcal{N}^{(r)}$, we have

$$\mathbf{p}^{(r)} = \mathbf{p}. \tag{5.5}$$

This follows directly from the fact that reversing time does not alter the average fraction of time the process spends in a state. Setting $-t-dt =: t_0$, the defining equation

$$\mathbb{P}(N_{t+dt} = \mathbf{n}, N_t = \mathbf{m}) = \mathbb{P}(N^{(r)}_{-t-dt} = \mathbf{n}, N^{(r)}_{-t} = \mathbf{m})$$
$$= \mathbb{P}(N^{(r)}_{t_0+dt} = \mathbf{m}, N^{(r)}_{t_0} = \mathbf{n})$$

leads to $p_\mathbf{m} \cdot \mathbb{P}(N_{t+dt} = \mathbf{n} \mid N_t = \mathbf{m}) = p_\mathbf{n}^{(r)} \cdot \mathbb{P}(N^{(r)}_{t_0+dt} = \mathbf{m} \mid N^{(r)}_{t_0} = \mathbf{n})$, such that, by dividing both sides by dt, letting $dt \to 0$, and observing (5.5), we obtain

$$p_\mathbf{m} \cdot g_\mathbf{mn} = p_\mathbf{n} \cdot g^{(r)}_\mathbf{nm}. \tag{5.6}$$

An important statement that characterizes the transition rates of the reversed process is the following.

Theorem 5.1 *Let* $\mathcal{N} = (N_t)_{t\in\mathbb{R}_0^+}$ *be a stationary Markov process with state space* E *and generator* $G = (g_\mathbf{mn})_{\mathbf{m},\mathbf{n}\in E}$. *Assume that there are nonnegative numbers* $g^*_\mathbf{mn}$ *satisfying*

$$\sum_{\mathbf{n}\in E} g_\mathbf{mn} = \sum_{\mathbf{n}\in E} g^*_\mathbf{mn} = 0 \ \text{for all} \ \mathbf{m} \in E,$$

and positive numbers $p_\mathbf{n}$, $\mathbf{n} \in E$, *summing to 1, such that the equations*

$$p_\mathbf{m} g_\mathbf{mn} = p_\mathbf{n} g^*_\mathbf{nm} \ \text{for all} \ \mathbf{m} \in E$$

are satisfied. Then $(g^*_\mathbf{mn})_{\mathbf{m},\mathbf{n}\in E} = G^{(r)}$ *is the generator of the reversed process, and the* $p_\mathbf{n}$, $\mathbf{n} \in E$, *form the stationary probability vector* $\mathbf{p}$ *for both, the reversed and the forward process.*

Proof: In order to show that $\mathbf{p} = (p_\mathbf{n})_{\mathbf{n}\in E}$ is the stationary vector of $\mathcal{N}$, observe that $\sum_{\mathbf{m}\in E} p_\mathbf{m} g_\mathbf{mn} = p_\mathbf{n} \sum_{\mathbf{m}\in E} g^*_\mathbf{nm} \overset{!}{=} p_\mathbf{n} \sum_{\mathbf{m}\in E} g_\mathbf{nm} = \mathbf{o}$, saying that $\mathbf{p}$ satisfies the global balance equation. Additionally, $p_\mathbf{m} g_\mathbf{mn} = p_\mathbf{n} g^*_\mathbf{nm}$ implies $g^*_\mathbf{nm} = g^{(r)}_\mathbf{nm}$ according to (5.6).
$\square$

Joint distributions of the original and the reversed process are not identical in general. With $t_1 < \ldots < t_k$, a kth-order joint distribution of $\mathcal{N}$ reads

$$p_{\mathbf{m}_1\ldots\mathbf{m}_k}(t_1,\ldots,t_k) = \mathbb{P}(N_{t_1} = \mathbf{m}_1,\ldots,N_{t_k} = \mathbf{m}_k),$$

whereas

$$\mathbb{P}(N_{t_1}^{(r)} = \mathbf{m}_1, \ldots, N_{t_k}^{(r)} = \mathbf{m}_k) = \mathbb{P}(N_{-t_1} = \mathbf{m}_1, \ldots, N_{-t_k} = \mathbf{m}_k),$$

which need not be the same as $p_{\mathbf{m}_1 \ldots \mathbf{m}_k}(t_1, \ldots, t_k)$.

Definition 5.2 A stochastic process is called *reversible*, if the joint distributions of the forward and the reversed process are identical, i.e.

$$\mathbb{P}(N_{t_1} = \mathbf{m}_1, \ldots, N_{t_k} = \mathbf{m}_k) = \mathbb{P}(N_{t_1}^{(r)} = \mathbf{m}_1, \ldots, N_{t_k}^{(r)} = \mathbf{m}_k).$$

Reversibility is related to the notion of detailed balance equations (5.4). First note that any reversible Markov process is stationary, as can immediately be deduced from the equality

$$\mathbb{P}(N_{t_1}, \ldots, N_{t_k}) = \mathbb{P}(N_{t_1+\tau}, \ldots, N_{t_k+\tau}) = \mathbb{P}(N_{-t_1}, \ldots, N_{-t_k})$$

for any $\tau \in \mathbb{R}_0^+$. Secondly, the following more general statement holds.

Theorem 5.3 *A stationary Markov process* $\mathcal{N} = (N_t)_{t \in \mathbb{R}_0^+}$ *is reversible if and only if the detailed balance equations (5.4) are satisfied for all* $\mathbf{m}, \mathbf{n} \in E$ *and some positive vector* $\mathbf{p} = (p_\mathbf{n})_{\mathbf{n} \in E}$ *with* $\sum_{\mathbf{n} \in E} p_\mathbf{n} = 1$.

Proof: 1. The properties of stationarity and reversibility imply that $\mathbb{P}(N_t = \mathbf{n})$ does not depend on t. The numbers $\mathbb{P}(N_t = \mathbf{n}) =: p_\mathbf{n}$ are positive and sum to 1. From reversibility and time-homogeneity we can conclude that

$$\mathbb{P}(N_{t+dt} = \mathbf{n}, N_t = \mathbf{m}) = \mathbb{P}(N_{t+dt}^{(r)} = \mathbf{n}, N_t^{(r)} = \mathbf{m})$$
$$= \mathbb{P}(N_{-t-dt} = \mathbf{n}, N_{-t} = \mathbf{m}),$$

which (setting $-t - dt =: t_0$) is equivalent to $p_\mathbf{m} \cdot \mathbb{P}(N_{t+dt} = \mathbf{n} \mid N_t = \mathbf{m}) = p_\mathbf{n} \cdot \mathbb{P}(N_{t_0+dt} = \mathbf{m} \mid N_{t_0} = \mathbf{n})$.[1] Forming the differential quotient on each side, one obtains (5.4).

2. The detailed balance equations guarantee global balance, so $\mathbf{p}$ represents the equilibrium distribution of the Markov process. Considering now an arbitrary interval $[-T, T]$, we calculate the joint probability density for the event that the process is in state $\mathbf{m}_1$ at time $-T$, jumps to state $\mathbf{m}_2$ at time $-T + x_1$, to state $\mathbf{m}_3$ at time $-T + x_1 + x_2$, and so forth, until it reaches state $\mathbf{m}_k$ at time $-T + \sum_{\nu=1}^{k-1} x_\nu$, staying there until T, i. e. for some time interval of length x_k that satisfies $\sum_{\nu=1}^{k} x_\nu = 2T$. The probability, that upon leaving a

---

[1]Remember, that it is even possible here to replace t_0 by t due to time-homogeneity.

state $\mathbf{m}_\nu$ the process jumps to state $\mathbf{m}_{\nu+1}$, is $g_{\mathbf{m}_\nu \mathbf{m}_{\nu+1}}/\gamma_{\mathbf{m}_\nu}$. Further, since we have a Markov process, the probability density of the sojourn time in state $\mathbf{m}_\nu$ equals $\gamma_{\mathbf{m}_\nu} e^{-\gamma_{\mathbf{m}_\nu} \cdot x_\nu}$, whereas we have $\mathbb{P}(\text{sojourn time in state } \mathbf{m}_k > x_k) = e^{-\gamma_{\mathbf{m}_k} \cdot x_k}$. As a consequence, the probability density for the above mentioned process behaviour in $[-T, T]$ reads

$$p_{\mathbf{m}_1} e^{-\gamma_{\mathbf{m}_1} \cdot x_1} g_{\mathbf{m}_1 \mathbf{m}_2} e^{-\gamma_{\mathbf{m}_2} \cdot x_2} g_{\mathbf{m}_2 \mathbf{m}_3} \cdots e^{-\gamma_{\mathbf{m}_{k-1}} \cdot x_{k-1}} g_{\mathbf{m}_{k-1} \mathbf{m}_k} e^{-\gamma_{\mathbf{m}_k} \cdot x_k}.$$

Applying now the detailed balance property, we obtain for the same density the expression

$$p_{\mathbf{m}_k} e^{-\gamma_{\mathbf{m}_1} \cdot x_1} g_{\mathbf{m}_k \mathbf{m}_{k-1}} e^{-\gamma_{\mathbf{m}_2} \cdot x_2} g_{\mathbf{m}_{k-1} \mathbf{m}_{k-2}} \cdots e^{-\gamma_{\mathbf{m}_{k-1}} \cdot x_{k-1}} g_{\mathbf{m}_2 \mathbf{m}_1} e^{-\gamma_{\mathbf{m}_k} \cdot x_k}$$

(since $p_{\mathbf{m}_1} g_{\mathbf{m}_1 \mathbf{m}_2} g_{\mathbf{m}_2 \mathbf{m}_3} = p_{\mathbf{m}_2} g_{\mathbf{m}_2 \mathbf{m}_1} g_{\mathbf{m}_2 \mathbf{m}_3} \overset{!}{=} p_{\mathbf{m}_3} g_{\mathbf{m}_3 \mathbf{m}_2} g_{\mathbf{m}_2 \mathbf{m}_1}$, etc.). This density, but, describes a process behaviour, where the process starts at time $-T$ in state $\mathbf{m}_k$, stays there for some time x_k, then jumps to state $\mathbf{m}_{k-1}$, stays there for some time x_{k-1}, and so forth, until it reaches state $\mathbf{m}_1$, where it remains at least for a period of x_1 time units. Consequently, the reversed process $(N_t)_{t \in [-T,T]}$ proves to behave exactly in the same way as the reversed process $(N_{-t})_{t \in [-T,T]}$. Since T has been arbitrarily chosen, $\mathcal{N}$ must be reversible.
□

For queueing network analyses the property of product form related to the state probabilities of isolated stations is of paramount importance. The following result relates reversibility with some other type of product form.

Theorem 5.4 *The stationary distribution of any irreducible and reversible Markov process can be calculated from a product of ratios of transition rates.*

Proof: Let $\mathcal{N}$ be an irreducible and reversible stationary Markov process, such that, according to theorem 5.3, the detailed balance equations (5.4) are satisfied. We select an arbitrary state, say $\mathbf{s} = (s_1, \ldots, s_M)$, as a fixed "starting state", from which any other state $\mathbf{n}$ is reachable due to irreducibility, that is, there is at least one sequence $\mathbf{s} = \mathbf{m}_1, \mathbf{m}_2, \ldots, \mathbf{m}_k = \mathbf{n}$ such that

$$g_{\mathbf{m}_1 \mathbf{m}_2} g_{\mathbf{m}_2 \mathbf{m}_3} \cdots g_{\mathbf{m}_{k-1} \mathbf{m}_k} \neq 0.$$

Using this fact, we further select for each state $\mathbf{n} \in E$ one and only one connecting sequence of this form with $\mathbf{m}_1 = \mathbf{s}$ and $\mathbf{m}_k = \mathbf{n}$, and define positive numbers $\pi_\mathbf{n}$ by

$$\pi_\mathbf{n} = \begin{cases} \dfrac{g_{\mathbf{m}_1 \mathbf{m}_2} g_{\mathbf{m}_2 \mathbf{m}_3} \cdots g_{\mathbf{m}_{k-1} \mathbf{m}_k}}{g_{\mathbf{m}_k \mathbf{m}_{k-1}} g_{\mathbf{m}_{k-1} \mathbf{m}_{k-2}} \cdots g_{\mathbf{m}_2 \mathbf{m}_1}} & \text{for } \mathbf{n} \neq \mathbf{s} \\ 1 & \text{for } \mathbf{n} = \mathbf{s} \end{cases}$$

(clearly, the intermediate states $\mathbf{m}_\nu$ as well as the value of the index k depend on $\mathbf{n}$). Next, setting $\sum_{\mathbf{n}\in E}\pi_{\mathbf{n}} = C$, we show that the distribution vector $\tilde{\mathbf{p}} = (\tilde{p}_{\mathbf{n}})_{\mathbf{n}\in E}$ with

$$\tilde{p}_{\mathbf{n}} \overset{!}{=} \frac{1}{C}\pi_{\mathbf{n}} = \frac{1}{C}\prod_{\nu=1}^{k(\mathbf{n})-1}\frac{g_{\mathbf{m}_\nu\mathbf{m}_{\nu+1}}}{g_{\mathbf{m}_{\nu+1}\mathbf{m}_\nu}}$$

satisfies the global balance equation (5.2). For that purpose, observe that according to detailed balance,

$$\pi_{\mathbf{n}} = \prod_{\nu=1}^{k(\mathbf{n})-1}\frac{g_{\mathbf{m}_\nu\mathbf{m}_{\nu+1}}}{g_{\mathbf{m}_{\nu+1}\mathbf{m}_\nu}} = \prod_{\nu=1}^{k(\mathbf{n})-1}\frac{p_{\mathbf{m}_{\nu+1}}}{p_{\mathbf{m}_\nu}} = \frac{p_{\mathbf{n}}}{p_{\mathbf{s}}},$$

which is true also for $\mathbf{n} = \mathbf{s}$. Consequently,

$$\sum_{\mathbf{n}\in E}\tilde{p}_{\mathbf{n}}g_{\mathbf{nm}} = \frac{1}{C}\left(g_{\mathbf{sm}} + \sum_{\substack{\mathbf{n}\in E \\ \mathbf{n}\neq\mathbf{s}}}\frac{p_{\mathbf{n}}}{p_{\mathbf{s}}}\cdot g_{\mathbf{nm}}\right) = \frac{1}{p_{\mathbf{s}}\,C}\sum_{\mathbf{n}\in E}p_{\mathbf{n}}\,g_{\mathbf{nm}} = 0,$$

implying that $\tilde{\mathbf{p}} = \mathbf{p}$. This proves the assertion.
□

Let $N_{i,t}$ denote the random number of customers in a single queueing station i at time t. If $(N_{i,t})_{t\in\mathbb{R}_0^+}$ is a stationary reversible Markov process then we call i a reversible queueing station. An important consequence from reversibility is the so-called **input-output property**: For any reversible queueing station the departure process has the same joint distribution as the arrival process. This is due to the fact that, whereas the points in time when $N_{i,t}$ increases by 1 correspond to arrivals, the points in time when $N_{i,t}$ decreases by 1 correspond to departures and, by definition, to the epochs when the reverse process $(N_{i,t}^{(r)})_{t\in\mathbb{R}_0^+}$ increases by 1. Since joint distributions of the original and the reverse process are the same, the arrival and the departure process exhibit the same joint statistics. As a consequence, we have the fact that a reversible queueing station, when being fed by a Poisson (Markov) stream, causes a Poisson (Markov) output stream. This property is called $M \Rightarrow M$ **property**.

Before considering other balance concepts, let us point to a general property of stationary Markov processes. Assume, as we do mostly in this chapter, that a stationary Markov process $\mathcal{N}$ can be interpreted as a random walk in a finite graph $\mathcal{G}$; then the rates of flow in opposite directions across a cut in $\mathcal{G}$ are identical. In other words, for some arbitrary subset A in the set E of nodes (the state space) we have[2]

$$\sum_{m\in A}\sum_{n\in E\backslash A} p_m\, g_{mn} = \sum_{m\in A}\sum_{n\in E\backslash A} p_n\, g_{nm}.$$

This is a direct consequence from global balance, since by summing on both sides of (5.3) over all n, and subtracting

$$\sum_{m\in A}\sum_{n\in E} p_m\, g_{mn} = \sum_{m\in A}\sum_{n\in A} p_n\, g_{nm}$$

, we obtain the above equation.

Opposed to the notions of global and detailed balance, the term **partial balance** plays an important role. In fact, the property of partial balance is the most general property, since global and detailed balance as well as other terms (such as station or local balance) can be regarded as special cases of partial balance.

An irreducible stationary Markov process with equilibrium distribution $p = (p_n)_{n\in E}$ and transition rates g_{mn} is said to be in **partial balance** with respect to a subset A of its state space E, if

$$\sum_{m\in A} p_m\, g_{mn} = \sum_{m\in A} p_n\, g_{nm}, \quad n\in A. \tag{5.7}$$

Notice, that the stationary distribution p satisfies the partial balance equations (5.7) if and only if

$$\sum_{m\in E\backslash A} p_m\, g_{mn} = \sum_{m\in E\backslash A} p_n\, g_{nm}, \quad n\in A;$$

this follows from stationarity, i. e. the fact that the process is in global balance and satisfies (5.3).

In many application oriented publications the property of partial balance is described in somewhat vague terms, e. g. saying that partial balance for some state m is present if the rate of flow out of m *due to changes of a particular nature* equals the rate of flow into m *due to changes of that very particular nature*.

It is here the point to pay attention to the fact that a network state, in general, is determined by several actual values of system parameters, rather than just by

---

[2] The exterior of a queueing network is represented by a node "0", such that q_{0i} is a routing probability into node i from outside the network, and q_{j0} is the routing probability from node j to outside the network. The node 0 is contained in the node set of $\mathcal{G}$.

the number of customers in each station (the latter definition leading to $E = \mathbb{N}_0^M$). For example, in a multi-class network with R classes a state description may contain information about the number of class r customers, their actual waiting (and/or server occupancy) positions, and the actual phases of service at every station $i \in \{1, \ldots, M\}$. Accordingly, the term partial balance includes a variety of specific definitions, among which the notions of *station balance* and *local balance* deserve particular notice.

Probably the most important property is that of *local balance*. Introduced by Chandy et alii [26], this term depicts a situation, where the rate of flow into a network state $\mathbf{m}$ due to the arrival of a class r customer at a network queue i is balanced against the rate of flow out of the same network state due to the departure of a class r customer from that network queue i. If the state description contains information about the actual phase of service in case of non-exponentially distributed service times, state changes are caused also by phase transitions or by an entry into the first phase of a service time distribution. Local balance, then, means that the probability flux into network state $\mathbf{m}$ due to the arrival of a class r customer at a network queue i by entering a service phase ℓ equals the probability flux out of the same network state due to the departure of a class r customer from that service phase ℓ at queue i. Chandy used the term "stage of service" for the tripel (i, r, ℓ) of queue index i, class index r, and phase index ℓ. Thus, a network is said to be in **local balance**, if the rate of flow into a stage (i, r, ℓ) of service is equal to the rate of flow out of the same stage (i, r, ℓ) of service for all admissible values of $i, r,$ and ℓ.

Let us write $g_{\mathbf{mn}}^{dep_i(r)}$ for the rate out of state $\mathbf{m}$ due to a departure of a class r customer from queue i (this rate is zero if there is no such customer at i in state $\mathbf{m}$), and $g_{\mathbf{nm}}^{arr_i(r)}$ for the rate into state $\mathbf{m}$ due to an arrival of a class r customer at queue i. The **local balance equations** then read

$$p_{\mathbf{n}} \, g_{\mathbf{nm}}^{arr_i(r)} = p_{\mathbf{m}} \, g_{\mathbf{mn}}^{dep_i(r)} \quad \text{for all} \quad i \in \{1, \ldots, M\}, \ 1 \le r \le R. \quad (5.8)$$

To illustrate the concept, consider a single class queueing network with M stations whose states are completely described by the vectors $\mathbf{m} = (m_1, \ldots, m_M)$ of station specific customer numbers. Let $\lambda_i(m_i)$ and $\mu_i(m_i)$, respectively, denote the arrival rate into, and the service completion rate at station i, when there are m_i customers present ($i = 1, \ldots, M$). According to the above definition of local balance, the rate of flow into some network state $\mathbf{m}$ due to an arrival at queue i must be equal to the rate of flow out of state $\mathbf{m}$ due to a departure from queue i. Let $\mathbf{e}_i$ denote a vector of length M that has a 1 at position i and zeros at all other positions (the ith canonical row base vector). An arrival at queue i can transfer a state $\mathbf{n}$ into $\mathbf{m}$ only if $\mathbf{n}$ equals $\mathbf{m} - \mathbf{e}_i$ (notice that a transition from $\mathbf{m} - \mathbf{e}_i + \mathbf{e}_j$ to $\mathbf{m}$ would be due to a departure from queue j, rather than due to an arrival at queue i). Similarly, a departure from queue i can transfer the state $\mathbf{m}$ only to one of the two states $\mathbf{m} - \mathbf{e}_i$ or $\mathbf{m} - \mathbf{e}_i + \mathbf{e}_j$. As a consequence, the local balance equations for that simple single class network with state space $\mathbb{N}_0^M$ read

$$p_{\mathbf{m}-\mathbf{e}_i} \lambda_i(m_i - 1) = p_{\mathbf{m}} \mu_i(m_i) q_{i0} + \sum_{j=1}^{M} p_{\mathbf{m}} \mu_i(m_i) q_{ij}$$

for all $i \in \{1, \ldots, M\}$, where q_{i0} and q_{ij}, respectively, are the routing probabilities from station i to the exterior of the network and to station j. Observing $q_{i0} + \sum_{i=1}^{M} q_{ij} = 1$ and $\mu_i(0) = 0$, we finally state that local balance means

$$p_{\mathbf{m}-\mathbf{e}_i} \lambda_i(m_i - 1) = p_{\mathbf{m}} \mu_i(m_i) \quad \text{if } m_i \geq 1, \quad i = 1, \ldots, M. \tag{5.9}$$

The next theorem should be considered as the central result with respect to the notion of local balance. We provide an exemplary proof only for the most simple situation of a single class network with state space $E = \mathbb{N}_0^M$, whose state descriptions $\mathbf{m} = (m_1, \ldots, m_M)$ reflect the station occupancies and whose routing probabilities are state independent. The general case can be handled similarly, although leading to more complex and intricate expressions. We refer to the books of Kelly [48], Kant [45] and Nelson [61] for further details.

Theorem 5.5 *Local balance implies global balance and product form.*

Proof: (For the most simple case with only one class of customer and state independent routing, where a state at some arbitrary point in time is determined by the actual numbers of customers present in the network queues.)

1. Assume that a probability distribution $(p_{\mathbf{n}})_{\mathbf{n} \in E}$ satisfies the local balance equations (5.8). We show that $(p_{\mathbf{n}})_{\mathbf{n} \in E}$ then satisfies global balance. Consider all neighbouring states $\mathbf{m} \pm \mathbf{e}_i$ and $\mathbf{m} \pm \mathbf{e}_i \mp \mathbf{e}_j$ that are reachable from state $\mathbf{m}$, such that the probability flux into state $\mathbf{m}$ is given by

$$\sum_{\substack{\mathbf{n} \in E \\ \mathbf{n} \neq \mathbf{m}}} p_{\mathbf{n}} g_{\mathbf{n}\mathbf{m}} = \sum_{i=1}^{M} p_{\mathbf{m}-\mathbf{e}_i} g_{\mathbf{m}-\mathbf{e}_i \, \mathbf{m}}^{arr_i}$$

$$+ \sum_{i=1}^{M} p_{\mathbf{m}+\mathbf{e}_i} g_{\mathbf{m}+\mathbf{e}_i \, \mathbf{m}}^{dep_i} + \sum_{i=1}^{M} \sum_{j=1}^{M} p_{\mathbf{m}+\mathbf{e}_i-\mathbf{e}_j} g_{\mathbf{m}+\mathbf{e}_i-\mathbf{e}_j \, \mathbf{m}}^{dep_i}.$$

Applying (5.8), this yields

$$\sum_{\substack{n \in E \\ n \neq m}} p_n g_{nm} = \sum_{i=1}^{M} p_m g_{m\,m-e_i}^{dep_i} + \sum_{i=1}^{M} \sum_{j=1}^{M} p_m g_{m\,m-e_i+e_j}^{dep_i} +$$

$$+ \sum_{i=1}^{M} p_{m+e_i} g_{m+e_i\,m}^{dep_i} + \sum_{i=1}^{M} \sum_{j=1}^{M} p_{m+e_i-e_j} g_{m+e_i-e_j\,m}^{dep_i}.$$

Now express m as $m = n - e_i$ in the first of the two sums in the second line, and as $m = n - e_i + e_j$ in the second one. Then these expressions can be rewritten as

$$\sum_{i=1}^{M} p_n g_{n\,n-e_i}^{dep_i} + \sum_{i=1}^{M} \sum_{j=1}^{M} p_n g_{n\,n-e_i+e_j}^{dep_i} \overset{!}{=} \sum_{i=1}^{M} p_{n-e_i} g_{n-e_i\,n}^{arr_i}$$

$$= \sum_{i=1}^{M} p_m g_{m\,m+e_i}^{arr_i},$$

such that the global flux into state m reads

$$\sum_{\substack{n \in E \\ n \neq m}} p_n g_{nm} = p_m \left\{ \sum_{i=1}^{M} g_{m\,m-e_i}^{dep_i} + \sum_{i=1}^{M} \sum_{j=1}^{M} g_{m\,m-e_i+e_j}^{dep_i} + \sum_{i=1}^{M} g_{m\,m+e_i}^{arr_i} \right\}.$$

The right hand side, but, of this expression is nothing else than the total probability flux out of state m, which proves that $(p_n)_{n \in E}$ satisfies global balance and, therefore, is the equilibrium state distribution of the network process $\mathcal{N}$. Consequently, all probability distributions over E that satisfy the local balance equations must coincide with the unique equilibrium state distribution of the network process $\mathcal{N}$.

2. We show that the distribution vector $(p_n)_{n \in E}$ that satisfies local balance has product form. Take any network station i in isolation, i.e. decoupled from the network, and provide the same input flow to i that the station experiences when communicationg with other stations in the network, such that the arrival and departure rates are the same as before. Obviously, local balance implies that i is in equilibrium. Let $p_i(m_i)$ be the steady state probability for the isolated station i to be in state m_i, and let $\lambda_i(m_i)$ and $\mu_i(m_i)$ denote the arrival rate into i and the departure rate from i, respectively, when i is in state m_i. Then, equations (5.8) take the form

$$p_i(m_i - 1)\lambda_i(m_i - 1) = p_i(m_i)\,\mu_i(m_i) \ \ \forall \ m_i \geq 1.$$

Define a probability vector $\tilde{\mathbf{p}} = (\tilde{p}_{\mathbf{n}})_{\mathbf{n} \in E}$ over $E = \mathbb{N}_0^M$ by

$$\tilde{p}_{\mathbf{n}} = \frac{1}{C_M} \prod_{i=1}^{M} p_i(n_i) \text{ for } \mathbf{n} = (n_1, \ldots, n_M),$$

where $C_M = \sum_{\mathbf{n} \in E} \prod_{i=1}^{M} p_i(n_i)$. In the network, the arrival rates into station i and the departure rates from station i, respectively, are

$$g_{\mathbf{m}-\mathbf{e}_i\,\mathbf{m}}^{arr_i} = \lambda_i(m_i - 1),$$
$$g_{\mathbf{m}\,\mathbf{m}-\mathbf{e}_i}^{dep_i} = \mu_i(m_i)\, q_{i0}, \quad g_{\mathbf{m}\,\mathbf{m}-\mathbf{e}_i+\mathbf{e}_j}^{dep_i} = \mu_i(m_i)\, q_{ij}$$

for $i, j \in \{1, \ldots, M\}$. The construction of $\tilde{\mathbf{p}}$ leads to

$$\tilde{p}_{\mathbf{m}-\mathbf{e}_i}\, g_{\mathbf{m}-\mathbf{e}_i\,\mathbf{m}}^{arr_i} = \frac{p_i(m_i - 1)}{p_i(m_i)} \lambda_i(m_i - 1),$$

$$\tilde{p}_{\mathbf{m}}\left(g_{\mathbf{m}\,\mathbf{m}-\mathbf{e}_i}^{dep_i} + \sum_{j=1}^{M} g_{\mathbf{m}\,\mathbf{m}-\mathbf{e}_i+\mathbf{e}_j}^{dep_i}\right) = \tilde{p}_{\mathbf{m}}\left(\mu_i(m_i)\, q_{i0} + \sum_{j=1}^{M} \mu_i(m_i)\, q_{ij}\right),$$

which implies that $\tilde{p}_{\mathbf{m}-\mathbf{e}_i}\, g_{\mathbf{m}-\mathbf{e}_i\,\mathbf{m}}^{arr_i} = \tilde{p}_{\mathbf{m}}\left(g_{\mathbf{m}\,\mathbf{m}-\mathbf{e}_i}^{dep_i} + \sum_{j=1}^{M} g_{\mathbf{m}\,\mathbf{m}-\mathbf{e}_i+\mathbf{e}_j}^{dep_i}\right)$, i.e. $\tilde{\mathbf{p}}$ satisfies the local balance equations. Consequently, $\tilde{\mathbf{p}}$ coincides with the uniquely determined equilibrium distribution $\mathbf{p} = (p_{\mathbf{n}})_{\mathbf{n} \in E}$ of $\mathcal{N}$.
$\square$

In general, it is necessary to be careful when reading statements on local balance in the literature since, unfortunately, there are no uniform standards for the definition of this notion. The reader who is interested in physical meanings and practice oriented versions is referred to the book of Van Dijk [30].

Another remark is in place addressing the property of *station balance*. Here the term "station" *does not* stand for "network station" in the sense of "a queue in the network" rather, it marks a position in the waiting or server room of a single queue that is occupied by one customer! A queue, in turn, is viewed as a set of stations. To illustrate the situation, consider an isolated multiple server queue that is visited by customers from different classes. Obviously, for a "first-come first-served" (FCFS) or "last-come first-served" (LCFS) scheduling discipline, the waiting positions at the top and the end of the queue, respectively, have particular meanings. Additionally, in case that specific servers are associated with specific classes, also the discrimination between servers (where service completions are to be expected) may be of importance. In that case to each server there is assigned a special subqueue containing customers at their waiting positions.

Bearing these peculiarities in mind, a "station" is determined by a position index j. A network queue i is viewed as a set of stations, and if there are in total n_i customers resident at i, the occupied stations are indexed $r_1, \ldots, r_{n_i}$, with r_j indicating the class of the customer at station (position) j. Even more complex descriptions are in use when routing chains are to be distinguished in the network, each containing users of different classes; we shall mention that below. In the more simple case, when discriminating between classes only, a possible single queue state definition is given by a $2n_i$ - tupel $\mathbf{n}_i := (r_1, \ldots, r_{n_i}, x_1, \ldots, x_{n_i})$, where i marks the queue in the network, n_i is the actual number of customers at this queue i, $\sigma_i = (r_1, \ldots, r_{n_i})$ forms the sequence of customer classes at positions $1, \ldots, n_i$, and $x_1, \ldots, x_{n_i}$ is a vector of remaining service requirements at these n_i positions. So, we have

$$\mathbf{m} = (\mathbf{n}_1, \ldots, \mathbf{n}_M)$$

when speaking of state $\mathbf{m} \in E$. A queueing network is said to be in **"station" balance** if during state $\mathbf{m}$ for any position ("station") j the actual fraction of the service rate associated with that position is proportional to the probability that a customer of the same category will arrive and be placed into this position. "Station" balance is tightly connected to the notion of symmetric service disciplines that we shall deal with in section 3 below. There we shall give a more precise definition. Clearly, "station" balance implies local balance and, consequently, global balance and product form. We have set the word "station" in quotation marks for two reasons: First, the term *position* in most cases reflects more precisely what is meant when describing a specific network state in a system with position depending dynamics and several chains and/or customer classes. Second, we wish to reserve the term *station* in this introductory book for a true queueing station in a network. There is a multitude of excellent books on that topic, and for details we refer to the literature mentioned at the end of this chapter.

Let us now turn back to the relationships between reversibility properties and flow balance. Asking for a property that guarantees partial balance we are led to the notion of quasi-reversibility. Let again $\mathcal{N} = (N_t : t \geq 0)$ be a Markov process with state space E that describes the dynamics of a queueing system serving customers from R different classes. As we saw already, a state $\mathbf{n} \in E$ may be identified by a fairly complex description, rather than merely by indicating the respective numbers of class specific customers in various stations.

Definition 5.6 $\mathcal{N}$ is called **quasi-reversible** if for any time t_0 the state N_{t_0} is independent of arrival times of class r customers after t_0 and departure times of class r customers prior to t_0 ($1 \leq r \leq R$).

A quasi-reversible process, in general, is not reversible (see exercise 5.3), and reversibility, in turn, does not imply quasi-reversibility. Accordingly, it should be stressed that these two notions are completely unrelated. For queueing networks, but, the property of quasi-reversibility is of significant pertinence. This is due to the fact that — as we shall see below — quasi-reversibility gives rise to product form expressions for the equilibrium state probabilities. Queues in "station" balance form an important subclass in the set of quasi-reversible queues. We first prove a result that is usually termed the *input-output property* of quasi-reversible queues.

Lemma 5.7 (Input-Output Property) *The arrival epochs as well as the departure epochs of class r customers in a stationary quasi-reversible queue form Poisson processes with class specific identical rates λ_r.*

Proof: The set of all states $\mathbf{n} \in E$ that provide the same state information as a given state $\mathbf{m}$ except that there is one class r customer more in the system, is marked $S(\mathbf{m} + r)$. Let $G = (g_{\mathbf{mn}})_{\mathbf{m,n} \in E}$ be the generator matrix of $\mathcal{N}$; then the rate of state changes due to class r arrivals when the state is $\mathbf{m}_t$ at time t is

$$\lambda_r(\mathbf{m}_t) = \sum_{\mathbf{n} \in S(\mathbf{m}_t + r)} g_{\mathbf{m}_t \mathbf{n}}.$$

1. According to quasi-reversibility the probability of a class r arrival during the interval $(t, t + dt]$ is independent of the state $\mathbf{m}_t$, and so is $\lambda_r(\mathbf{m}_t) = \lambda_r$. Further, according to the Markov property, the path realization prior to t has no influence on the probability for an arrival in $(t, t + dt]$, which means that the arrival process is memoryless with rate λ_r, independent of all earlier states prior to t. Consequently, the class r arrival epochs form an independent Poisson process with rate λ_r.
2. Interchanging the meaning of arrivals and departures of class r customers, the reverse process $\mathcal{N}^{(r)}$ again is to be interpreted as the state process of a queue with R customer classes, and since $\mathcal{N}$ is quasi-reversible, so is $\mathcal{N}^{(r)}$. Therefore, the same reasoning applies, stating that the class r arrival process of $\mathcal{N}^{(r)}$ forms a Poisson process with rate $\delta_r = \sum_{\mathbf{n} \in S(\mathbf{m}_t + r)} g^{(r)}_{\mathbf{mn}}$. This rate is the class r departure rate of $\mathcal{N}$, and so, due to stationarity, equals λ_r, which proves the assertion.
$\square$

We are now in the position to formulate the relationship between quasi-reversibility and partial balance.

Lemma 5.8 *Any quasi-reversible Markov process $\mathcal{N} = (N_t : t \geq 0)$ over some state space E that describes the dynamics of a queueing system with R*

customer classes satisfies partial balance with respect to the set $S(\mathbf{m} + r)$ for any $\mathbf{m} \in E$.

Proof: Remember, that $S(\mathbf{m} + r)$ describes the set of all states $\mathbf{n} \in E$ that provide the same state information as a given state $\mathbf{m}$ except that there is one class r customer more in the system. From equation (5.6) we obtain

$$p_\mathbf{m} \sum_{\mathbf{n} \in S(\mathbf{m}+r)} g_\mathbf{mn}^{(r)} = \sum_{\mathbf{n} \in S(\mathbf{m}+r)} p_\mathbf{n} \cdot g_\mathbf{nm},$$

since the reversal of the reverse process is the original one. According to the proof of lemma 5.7,

$$\lambda_r = \sum_{\mathbf{n} \in S(\mathbf{m}+r)} g_\mathbf{mn} = \sum_{\mathbf{n} \in S(\mathbf{m}+r)} g_\mathbf{mn}^{(r)},$$

and so

$$p_\mathbf{m} \sum_{\mathbf{n} \in S(\mathbf{m}+r)} g_\mathbf{mn} = \sum_{\mathbf{n} \in S(\mathbf{m}+r)} p_\mathbf{n} \cdot g_\mathbf{nm}.$$

$\square$

Let us now consider a vector-valued continuous time Markov process $\mathcal{N} = (N_t : t \geq 0)$ with state space E that describes a multi-class queueing network with M stations and R classes of customers. Upon completing service at one station i, a class r customer not only may join another station j of the network or depart from the network, but also may change its class before joining another queue. In general, the probability to undergo such type of change may depend on the history of the customer's behaviour and on the state of the process. The analysis of queueing networks of that generality has turned out to be very complex, if not impossible. When speaking of Markovian queueing networks in this chapter, we mean a subclass of networks that is characterized by the property that the routing probabilities are memoryless and independent of the network states, this way defining the transition matrix of a Markov chain.

Let $q_{ir;jr'}$ denote the probability that a class r customer, after leaving queue i, joins queue j as a class r' customer, and set $q_{ir;00}$ for the probability that a class r customer leaves the network after service completion at station i. Then $\sum_{j=1}^{M} \sum_{r'=1}^{R} q_{ir;jr'} + q_{ir;00} = 1$, and the discrete time - discrete state Markov chain defined by

$$Q = (q_{ir;jr'})_{i,j \in \{0,...,M\}, r,r' \in \{0,...,R\}}^{3}$$

---

[3] Where, for $j = 0$ only $r' = 0$ is possible, and vice versa.

is called the **routing chain**. A queueing network of that type is said to perform **Markov routing**.

Remark 5.9 If an asymptotic distribution **p** for a Markovian network process $\mathcal{N} = (N_t : t \geq 0)$ does exist, then there is an asymptotic marginal distribution $\mathbf{p}_i = p_i(\mathbf{k})$ for any queue i, too.

This can be seen from

$$p_i(\mathbf{k}) = \lim_{t\to\infty} \mathbb{P}(pr_i(N_t) = \mathbf{k}) = \lim_{t\to\infty} \sum_{\mathbf{n}\in pr_i^{-1}(\mathbf{k})} \mathbb{P}(N_t = \mathbf{n})$$

$$= \sum_{\mathbf{n}\in pr_i^{-1}(\mathbf{k})} \lim_{t\to\infty} \mathbb{P}(N_t = \mathbf{n}) = \sum_{\mathbf{n}\in pr_i^{-1}(\mathbf{k})} p_{\mathbf{n}},$$

with pr_i denoting the projection on the ith station specific state component, and $pr_i^{-1}(k) = \{\mathbf{n} \in \mathbb{N}_0^M : n_i = k\}$.

We close this section by formulating some sort of a quintessence from the above treatment of quasi-reversibility.

Theorem 5.10 *Let $\mathcal{N} = (N_t : t \geq 0)$ be a stationary vector-valued continuous time Markov process with state space E that describes a multi-class queueing network with M stations and R classes of customers. If each queueing station in isolation behaves as a quasi-reversible queue, and if the network performs Markov routing, then $\mathcal{N}$ is again quasi-reversible, and its equilibrium distribution $\mathbf{p} = (p_{\mathbf{n}})_{\mathbf{n}\in E}$ assumes product form, i.e.*

$$p_{\mathbf{n}} = \frac{1}{C} \prod_{i=1}^{M} f_i(\mathbf{n}_i),$$

where $f_i(\mathbf{n}_i)$ is a state depending function for an isolated station i in steady state $\mathbf{n}_i$, and C is some normalization factor.

We give a sketch of the proof for the simple case of a network whose states are defined by class specific customer occupancies only, and in which no class changes occur. For the more general cases we refer to the excellent treatments given by Kelly [61], and Nelson [48].

Proof: Stationarity of the whole network implies that of any single station. Consider a station i in isolation with the same class specific input streams, and let $p_i(\mathbf{n}_i) = p_i(k_{i1}, \ldots, k_{iR})$ for $k_r \in \mathbb{N}_0$ and $r \in \{1, \ldots, R\}$ be its steady state distribution. Quasi-reversibility means that the input and the output stream for each customer class r at i are Poisson (with same rate λ_{ir}), so we have $p_i(k_{i1}, \ldots, k_{ir} - 1, \ldots, k_{iR}) \lambda_{ir} = p_i(k_{i1}, \ldots, k_{iR}) \mu_{ir}(k_{ir})$. Construct a probability distribution by

$$p_{\mathbf{n}} = \frac{1}{C(M, R)} \prod_{i=1}^{M} p_i(k_{i1}, \ldots, k_{iR}).$$

Then this distribution satisfies local balance (cf. proof of theorem 5.5) and, therefore, also global balance.

$\square$

Notice, that the essential property here for a product form to hold is the property of each station to produce, when being fed by a Poisson input stream, an output stream that again is Poisson. This is nothing else than the $M \Rightarrow M$ property.

2.   Jackson and Gordon-Newell Networks

Let us consider now the simplest type of queueing network. This is an open or closed single class network of, say, M queues, whose state space is determined by the station specific numbers of customers only. Let $\mathcal{N} = (N_t : t \geq 0)$ denote the stochastic process that describes the dynamics of such a network with respect to the varying numbers of customers in the stations.[4] Its state space E is a subset of $\mathbb{N}_0^M$.

As before, we denote with $Q = (q_{ij})_{i,j \in \{1,\ldots,M\}}$ the routing matrix, and with $G = (g_{\mathbf{mn}})_{\mathbf{m},\mathbf{n} \in \mathbb{N}_0^M}$ the generator of $\mathcal{N}$. An open network of that kind is called a **Jackson network** if the following conditions are satisfied.

1   Any user entering $\mathcal{N}$ at some node i may reach any other node in finitely many steps with positive probability. Similarly, starting from some node i a user can leave the network in finitely many steps with positive probability ($i \in \{1, \ldots, M\}$).

2   The network performs Markov routing, i.e. Q represents the transition matrix of a Markov chain.

3   Each queueing station i is of type $*/M/s_i$ with $s_i \in \mathbb{N} \cup \{\infty\}$, i.e. the service time distribution at station i is exponential with parameter μ_i for each of the s_i servers.

---

[4]The letter $\mathcal{N}$ may also stand for the queueing network itself, as long as no ambiguities are to be expected.

4 The total arrival stream from outside the network forms a Poisson stream of intensity γ. The separate arrival streams to stations $i = 1, \ldots, M$ are determined by the routing probabilities q_{0i} with $\sum_{i=1}^{M} q_{0i} = 1$. They are Poisson streams with intensities $\gamma \, q_{0i} =: \gamma_i$.

$\mathcal{N}$ is irreducible due to property 1, and is Markov due to the memoryless property of the exponential service and inter-arrival time distributions. As such, the network assumes equilibrium if and only if there exists a positive finite solution $\mathbf{p} = (p_{\mathbf{n}})_{\mathbf{n} \in E}$ to the system of equations

$$\mathbf{p} \, G = \sum_{\mathbf{m} \in E} p_{\mathbf{m}} \, g_{\mathbf{mn}} = \mathbf{0},$$

where $G = (g_{\mathbf{mn}})_{\mathbf{m}, \mathbf{n} \in \mathbb{N}_0^M}$ is the generator of $\mathcal{N}$.

Let λ_i and δ_i, respectively, denote the total mean arrival and departure rates at stations $i = 1, \ldots, M$, each being independent of the actual occupancy at the stations. Then, $\lambda_i = \gamma_i + \sum_{j=1}^{M} \delta_j \, q_{ji}$. In equilibrium, $\delta_i = \lambda_i$ for each $i \in \{1, \ldots, M\}$, and so

$$\lambda_i = \gamma_i + \sum_{j=1}^{M} \lambda_j \, q_{ji}, \quad i = 1, \ldots, M. \tag{5.10}$$

(5.10) is called the **system of traffic equations** for a Jackson network in equilibrium. The next lemma shows that this system always possesses a unique solution.

Lemma 5.11 *For a Jackson network with routing matrix Q, the matrix $I - Q$ is invertible.*

Proof: Consider a Markov chain X with transition matrix

$$P = \begin{pmatrix} 1 & \mathbf{0} \\ \mathbf{q}_0 & Q \end{pmatrix},$$

where $\mathbf{q}_0 = (q_{10}, \ldots, q_{M0})^T$ is a column vector, and $\mathbf{0} = (0, \ldots, 0)$ is the zero row vector. Irreducibility of the Jackson network implies that the set of states $\{1, \ldots, M\}$ forms a transient communication class in the state space of X, whereas the state zero is absorbing. Hence, according to corollary 2.15, the submatrix $\tilde{R}$ of the potential matrix R of X (as defined in (2.7)) that contains entries $R(ij)$ with $i, j \in \{1, \ldots, M\}$ only, is finite. Due to the structure of P, $\tilde{R} = \sum_{n=1}^{\infty} Q^n$, and since $\tilde{R}$ is finite, the Neumann series

$$\tilde{R} + I = \sum_{n=0}^{\infty} Q^n = (I - Q)^{-1}$$

is finite, too. This proves the assertion.
□

We denote, as usual, by $\rho_i = \lambda_i/\mu_i$ the load factor of station i, $1 \leq i \leq M$. In general, the service completion rate at each station i is state dependent, given by $\mu_i(n_i) = \mu_i \min(s_i, n_i)$ when there are n_i customers present at i. Obviously, a necessary condition for the network process $\mathcal{N}$ to attain equilibrium is that all individual station specific processes attain equilibrium, i.e. stationarity of $\mathcal{N}$ implies

$$\rho_i < s_i \quad \text{for all} \quad i \in \{1, \ldots, M\}. \tag{5.11}$$

The following statement has first been proven by Jackson as early as in 1963 [42]. It shows that (5.11) not only is a necessary, but also a sufficient condition for stationarity of $\mathcal{N}$, and that any Jackson network is a product form (PF) network.

Theorem 5.12 (Jackson) *Let $\mathcal{N}$ denote a Markov process describing a Jackson network with M stations, and assume $\rho_i < s_i$ for all $1 \leq i \leq M$. Then a stationary distribution of $\mathcal{N}$ exists and is given by*

$$p_{\mathbf{n}} = \prod_{i=1}^{M} p_i(n_i), \tag{5.12}$$

for $\mathbf{n} = (n_1, \ldots, n_M)$, where $p_i = (p_i(n_i))_{n_i \in \mathbb{N}_0}$ is the stationary distribution of an isolated $M/M/s_i$ queueing system with arrival rate λ_i and service rate μ_i at each server.

Proof: $(p_{\mathbf{n}})_{\mathbf{n} \in \mathbb{N}_0^M}$ is a probability distribution since $p_i(n_i) \geq 0$ for all $i \in \{1, \ldots, M\}$, and

$$\sum_{\mathbf{n} \in \mathbb{N}_0^M} p_{\mathbf{n}} = \sum_{n_1=0}^{\infty} \cdots \sum_{n_M=0}^{\infty} \prod_{i=1}^{M} p_i(n_i) = \prod_{i=1}^{M} \sum_{n_i=0}^{\infty} p_i(n_i) = 1.$$

From equations (5.11), (5.12), and (5.9) we know that

$$p_{\mathbf{m}-\mathbf{e}_i} \lambda_i = p_{\mathbf{m}} \mu_i(s_i) \quad \text{for all} \quad i \in \{1, \ldots, M\},$$

which means that the distribution (5.9) satisfies local balance. Thus, by theorem 5.5, $(p_{\mathbf{n}})_{\mathbf{n} \in \mathbb{N}_0^M}$ is the equilibrium distribution of $\mathcal{N}$.
□

Jackson has proved this theorem by directly establishing the global balance relations. We repeat his rationale here for pedagogical reasons in order to

illustrate the interplay of input and output flows in a Jackson network.
For a Jackson network, the transition rates g_{nm} *into* a network state $\mathbf{m} = (m_1, \ldots, m_M)$ read

$$
g_{nm} = \begin{cases} \gamma_i & \text{if } \mathbf{n} = \mathbf{m} - \mathbf{e}_i \\ \mu_j(n_j + 1) \cdot q_{ji} & \text{if } \mathbf{n} = \mathbf{m} - \mathbf{e}_i + \mathbf{e}_j \\ \mu_i(n_i + 1) \cdot q_{i0} & \text{if } \mathbf{n} = \mathbf{m} + \mathbf{e}_i \end{cases} ,
$$

whereas the rates *out of* a network state $\mathbf{m} = (n_1 \ldots, n_M)$ read

$$
g_{mn} = \begin{cases} \gamma_i & \text{if } \mathbf{n} = \mathbf{m} + \mathbf{e}_i \\ \mu_i(n_i) \cdot q_{ij} & \text{if } \mathbf{n} = \mathbf{m} - \mathbf{e}_i + \mathbf{e}_j \\ \mu_i(n_i) \cdot q_{i0} & \text{if } \mathbf{n} = \mathbf{m} - \mathbf{e}_i \end{cases} .
$$

Due to $\rho_i < s_i$ for all i, each network station in isolation with same arrival rates assumes equilibrium, satisfying the local balance equations $p_i(m_i+1) \mu_i(m_i + 1) = p_i(m_i) \lambda_i$ for all $m_i \geq 0$, $1 \leq i \leq M$. Consequently, an expression of the form (5.12) leads to[5]

$$
p_{\mathbf{m}-\mathbf{e}_i} = \prod_{k \neq i} p_k(m_k)\, p_i(m_i - 1) = p_{\mathbf{m}} \frac{\mu_i(m_i)}{\lambda_i},
$$

$$
p_{\mathbf{m}-\mathbf{e}_i+\mathbf{e}_j} = \prod_{k \neq i,j} p_k(m_k)\, p_i(m_i - 1)\, p_j(m_j + 1) = p_{\mathbf{m}} \frac{\mu_i(m_i)}{\lambda_i} \frac{\lambda_j}{\mu_j(m_j + 1)},
$$

$$
p_{\mathbf{m}+\mathbf{e}_i} = \prod_{k \neq i} p_k(n_k)\, p_i(n_i + 1) = p_{\mathbf{m}} \frac{\lambda_i}{\mu_i(m_i + 1)},
$$

and the probability flow *into* network state $\mathbf{m}$ is

$$
\sum_{\mathbf{n} \in \mathbb{N}_0^M} p_{\mathbf{n}}\, g_{nm} = p_{\mathbf{m}} \left\{ \sum_{i=1}^M \frac{\mu_i(m_i)}{\lambda_i} + \sum_{i=1}^M \sum_{j=1}^M \frac{\mu_i(m_i)}{\lambda_i} \lambda_j\, q_{ji} + \sum_{i=1}^M \lambda_i\, q_{i0} \right\},
$$

which, according to $\sum_{j=1}^M \lambda_j\, q_{ji} = \lambda_i - \gamma_i$ (which follows from the traffic equations) and $\sum_{j=1}^M q_{ij} = 1 - q_{i0}$, reduces further to

$$
\sum_{\mathbf{n} \in \mathbb{N}_0^M} p_{\mathbf{n}}\, g_{nm} = p_{\mathbf{m}} \sum_{i=1}^M \Big(\mu_i(m_i) + \gamma_i \Big). \tag{5.13}
$$

---

[5]Note that $p_i(\nu) = 0$ for $\nu < 0$, and $\mu_i(\nu) = 0$ for $\nu \leq 0$, $1 \leq i \leq M$.

On the other side, by the same reasoning, the probability flow *out of* network state m can be rewritten as

$$\sum_{n \in \mathbb{N}_0^M} p_m \, g_{mn} = p_m \left\{ \sum_{i=1}^{M} \gamma_i + \sum_{i=1}^{M} \sum_{j=1}^{M} \mu_i(m_i) \, q_{ij} + \sum_{i=1}^{M} \mu_i(m_i) \, q_{i0} \right\},$$

and this, as is easily seen, is the same as (5.13), proving theorem 5.12.

A closed network possessing all the properties 1 - 3 of a Jackson network (with the exception of property 4) is called a **Gordon-Newell network,** or GN network for short. As shown by W. J. Gordon and G. F. Newell in 1967 [40], such a network assumes equilibrium with stationary distribution

$$p_n = \frac{1}{\tilde{C}_M(K)} \prod_{i=1}^{M} p_i(n_i) \tag{5.14}$$

for $\mathbf{n} = (n_1, \ldots, n_M)$, where again $p_i = (p_i(n_i))_{n_i \in \mathbb{N}_0}$ is the stationary distribution of an isolated $M/M/s_i$ queueing system with arrival rate λ_i and service rate μ_i at each server, and where $\tilde{C}_M(K) = \sum_{n \in E} \prod_{i=1}^{M} p_i(n_i)$ represents a normalization factor that guarantees $\sum_{n \in E} p_n = 1$ (K the constant number of customers in the network).

This statement is usually called the **Theorem of Gordon-Newell.** Its proof is given by the same reasoning as for the theorem of Jackson by setting $\gamma_i = 0$ and $q_{i0} = 0$ for $1 \leq i \leq M$. In both cases the participating network stations behave as if being completely independent, a result that is somewhat surprising, since — at least for a Gordon-Newell network — the dependency of station specific events is obvious: Given, that there are K customers in the network, we always have $\sum_{i=1}^{M} n_i = K$. The reason behind is the $M \Rightarrow M$ property that implies local balance.

The state space $E = E(M, K)$ of a Gordon-Newell network with M stations and K customers is given as the set of vectors

$$E(M, K) = \left\{ \mathbf{n} = (n_1, \ldots, n_M) : n_i \geq 0 \; \forall \, i, \; \sum_{i=1}^{M} n_i = K \right\}$$

and has size

$$|E(M, K)| = \binom{M + K - 1}{M - 1}.$$

The latter is easily seen by induction: Obviously, $|E(M, 0)| = |E(1, K)| = 1$. Further, we have $|E(2, K)| = K + 1$, since, according to $n_1 + n_2 = K$, any state $\mathbf{n} = (n_1, n_2)$ is already determined by only one of its entries $n_i \in$

$\{0, 1, \ldots, K\}$. Assume that $|E(M - 1, K)| = \binom{M+K-2}{M-2}$. Adding another node to the network that is appropriately connected with the former nodes, the new node may contain $\nu \in \{0, 1, \ldots, K\}$ users when there are $K - \nu$ users at the remaining $M - 1$ nodes. Hence,

$$
|E(M, K)| = \sum_{\nu=0}^{K} \binom{M + (K - \nu) - 2}{M - 2} = \sum_{\nu=0}^{K} \binom{M - 2 + \nu}{M - 2}
$$

$$
= \binom{M - 1}{M - 1} + \sum_{\nu=1}^{K} \binom{M - 2 + \nu}{M - 2},
$$

and the well known relation

$$
\binom{n}{k} + \binom{n}{k + 1} = \binom{n + 1}{k + 1},
$$

with $k = M - 2$ and $n = M - 2 + \nu$, yields

$$
|E(M, K)| = \binom{M - 1}{M - 1} + \sum_{\nu=1}^{K} \left(\binom{M - 1 + \nu}{M - 1} - \binom{M - 2 + \nu}{M - 1} \right)
$$

$$
= \binom{M - 1 + K}{M - 1}.
$$

2.1 The Performance of a Jackson Network

The performance measures of a Jackson network are easily obtained from those of isolated $M/M/s_i$ stations. As has previously been shown,

$$
p_i(0) = \left(\sum_{k=0}^{s_i-1} \frac{\rho_i^k}{k!} + \frac{\rho_i^{s_i}}{s_i!} \left(1 - \frac{\rho_i}{s_i} \right)^{-1} \right)^{-1},
$$

$$
p_i(n_i) = \begin{cases} p_i(0) \dfrac{\rho_i^{n_i}}{n_i!} & \text{for } 0 \le n_i \le s_i \\ p_i(0) \left(\dfrac{\rho_i}{s_i} \right)^{n_i} \dfrac{s_i^{s_i}}{s_i!} & \text{for } n_i \ge s_i \end{cases},
$$

where $\rho_i = \lambda_i/\mu_i$.[6] So, for a Jackson network, any state probability is immediately obtained from (5.12), whereas for a Gordon-Newell network it is necessary to additionally compute the normalization constant. Notice, that the steady state probabilities depend only on the mean values $\bar{x}_i = 1/\mu_i$ of the

---

[6]In some publications the quantity ρ_i is defined as $\rho_i = \lambda_i/(s_i \mu_i)$ and termed *utilization factor*; this is the mean fraction of active servers (see [50]).

service time distributions, and not on higher moments. This property is common to all product form networks, and is called the **product form network insensitivity property**.

Let, for an $M/M/s_i$ station, $\bar{N}_i$ denote the mean number of customers in the station, $\bar{T}_i$ the mean sojourn time, $\bar{W}_i^Q$ the mean waiting time in the queue, $\bar{N}_i^Q$ the mean queue length, and $\bar{S}_i$ the mean throughput through the station. Then,

$$\bar{N}_i = \rho_i + p_i(0)\frac{\rho_i^{s_i+1}}{(s_i-1)!(s_i-\rho_i)^2},$$

$$\bar{T}_i = \frac{1}{\mu_i}\left(1 + p_i(0)\frac{\rho_i^{s_i}}{(s_i-1)!(s_i-\rho_i)^2}\right),$$

$$\bar{W}_i^Q = \frac{1}{\mu_i}p_i(0)\frac{\rho_i^{s_i}}{(s_i-1)!(s_i-\rho_i)^2},$$

$$\bar{N}_i^Q = \bar{N}_i - \rho_i,$$

$$\bar{S}_i = \lambda_i,$$

and the total average number $\bar{N}$ of customers is

$$\bar{N} = \sum_{i=1}^{M} \bar{N}_i.$$

Applying Little's result (see theorem 1.9), the total mean sojourn time or network delay a customer experiences is obtained as

$$\bar{T} = \frac{1}{\gamma}\bar{N},$$

where $\gamma = \sum_{i=1}^{M} \gamma_i = \bar{S}$ is the total mean throughput through the network. We denote by τ_i the mean time between a user's arrival at node i and his final departure from the network. For this quantity we immediately realize the relation

$$\tau_i = \bar{T}_i + \sum_{j=1}^{M} q_{ij}\tau_j, \quad 1 \le i \le M.$$

Let v_i denote the mean number of visits a user makes at sation i. The total number of customers that enter the network per unit time is γ, and each of these customers visits node i in the average for v_i times, so γv_i gives the average ratio of arrivals per unit time at station i, implying that $\gamma v_i = \lambda_i$. The traffic equations (5.10), therefore, yield

$$v_i = \frac{\gamma_i}{\gamma} + \sum_{j=1}^{M} v_j q_{ij}, \quad 1 \le i \le M. \tag{5.15}$$

For any Jackson network the system (5.15) always possesses a unique non-negative solution due to lemma 5.11.

2.2 Computational Methods for Gordon-Newell Networks

The calculation of performance measures of a Gordon-Newell network is by far not as easy as in the case of Jackson networks. The main problem consists in computing the normalization constant

$$\tilde{C}_M(K) = \sum_{n \in E(M,K)} \prod_{i=1}^{M} p_i(n_i). \tag{5.16}$$

What is the reason? It is simply the fact that the huge number of terms occurring in (5.16) makes it very difficult, in general, to numerically evaluate the product form solution. Special algorithmic methods are in place here, and we shall demonstrate one below.

The Convolution Algorithm

Consider the traffic equations of a GN network,

$$\lambda_i = \sum_{j=1}^{M} \lambda_j\, q_{ij}, \quad 1 \leq i \leq M. \tag{5.17}$$

Obviously, the quantities λ_i are only determined up to some non-zero constant, and in order not to identify them with the "true" arrival rates, it is convenient to replace the term λ_i by y_i and just look at these y_i as solutions of the above system (5.17). For technical reasons we set

$$x_i(n_i) := \frac{p_i(n_i)}{p_i(0)}, \quad \text{and} \quad C_M(K) := \frac{\tilde{C}_M(K)}{\prod_{i=1}^{M} p_i(0)}.$$

The product form equation (5.14) then takes the form

$$p_\mathbf{n} = \frac{1}{C_M(K)} \prod_{i=1}^{M} x_i(n_i), \tag{5.18}$$

and according to the local balance equations $p_i(n_i + 1)\, \mu_i(n_i + 1) = p_i(n_i)\, \lambda_i$ as well as the convention $\lambda_i = y_i$ we have

$$x_i(0) = 1, \quad x_i(k) = x_i(k-1)\frac{y_i}{\mu_i(k)} \quad \text{for } 1 \leq k \leq K. \tag{5.19}$$

Thus, given any solution $y_1, \ldots, y_M$ of the system of traffic equations (5.17), we can compute all the $x_i(n_i)$ simply by iteration.

Let us pause here for a moment in order to introduce the notion of *discrete convolution* of vectors of equal length (or even sequences with infinitely many components): Given $\mathbf{a} = (a_1, \ldots, a_N)$ and $\mathbf{b} = (b_1, \ldots, b_N)$, the convolution of $\mathbf{a}$ and $\mathbf{b}$ is defined as the vector $\mathbf{c} = (c_1, \ldots, c_N)$ of same length that has the components

$$c_k = \sum_{\ell=0}^{k} a_{k-\ell}\, b_\ell = \sum_{\ell=0}^{k} a_\ell\, b_{k-\ell}, \quad 0 \le k \le N$$

($N \le \infty$). The common symbol for the convolution operation is the "$*$", i.e. we write $\mathbf{c} = \mathbf{a} * \mathbf{b}$. It is obvious that $(\mathbf{a} * \mathbf{b}) * \mathbf{c} = \mathbf{a} * (\mathbf{b} * \mathbf{c})$ for arbitrary vectors $\mathbf{a}, \mathbf{b}, \mathbf{c} \in \mathbb{R}^N$. For a convolution of some vector $\mathbf{a} \in \mathbb{R}^N$ with itself we write

$$\mathbf{a}^{*n} = \mathbf{a}^{*n-1} * \mathbf{a} \quad \text{for } n \ge 1,$$

where $\mathbf{a}^{*0}$ is defined as $\mathbf{a}^{*0} = (1, 0, \ldots, 0)$, hence $\mathbf{a}^{*0} * \mathbf{b} = \mathbf{b}$ for all $\mathbf{b} \in \mathbb{R}^N$.

We return now to the problem of computing the steady state probabilities (5.18). Although the $x_i(n_i)$ can easily be computed by iteration, the computation of the normalization constant $C_M(K) = \sum_{\mathbf{n} \in E_M(K)} \prod_{i=1}^{M} x_i(n_i)$ still turns out to be rather difficult if M and K attain large values. In this situation J. Buzen [24] observed that (5.16) is nothing else than the Kth component of the discrete convolution of the vectors

$$\mathbf{x}_i = (x_i(0), x_i(1), \ldots, x_i(K)).$$

Precisely, we have

$$C_M(K) = \sum_{\mathbf{n} \in E(M,K)} \prod_{i=1}^{M} x_i(n_i) \overset{!}{=} (\mathbf{x}_1 * \ldots * \mathbf{x}_M)(K). \tag{5.20}$$

Formally, expression (5.20) is characterized by the two parameters M and K, and so it is suggesting itself that we define, for $1 \le m \le M$, $1 \le k \le K$, the components $C_m(k)$ of the convolution vector $\mathbf{C}_m = \mathbf{x}_1 * \ldots * \mathbf{x}_m$ by

$$C_m(k) = \sum_{\mathbf{n} \in E(m,k)} \prod_{i=1}^{m} x_i(n_i) = (\mathbf{x}_1 * \ldots * \mathbf{x}_m)(k),$$

where $E(m, k) = \left\{ \mathbf{n} = (n_1, \ldots, n_m) : n_i \ge 0, \; \sum_{i=1}^{m} n_i = k \right\}$. Similarly, the constant $C_M(K)$ can be written as the Kth component of the convolution of $\mathbf{C}_{M-1}$ and $\mathbf{x}_M$:

$$C_M(K) = (\mathbf{C}_{M-1} * \mathbf{x}_M)(K).$$

In general terms, we arrive at

$$C_m(k) = (\mathbf{C}_{m-1} * \mathbf{x}_m)(k), \quad 1 \le m \le M, \ 1 \le k \le K. \quad (5.21)$$

This, in fact, is the basis of Buzen's convolution algorithm. It can roughly be described as follows.

1. Set $C_0(0) = 1$, and $C_0(\ell) = 0$ for $1 \le \ell \le K$.

2. For all m, $1 \le m \le M$, set $x_m(0) = 1$.

3. Compute successively, for any $m \in \{1, \dots, M\}$ and $k = 0, \dots, K$, the values $x_m(k) = x_{m-1} y_m/\mu_m(k)$ and $C_m(k) = \sum_{\mathbf{n} \in E_m(k)} \prod_{i=1}^{m} x_i(n_i)$.

Performance Measures

The computation of all the (normalization) constants $C_m(k)$ opens the way for an easy and direct evaluation of station specific performance measures. Note that, by adequate renumbering, we always can achieve that an arbitrary station has index M. Let $p_M(n; K)$ denote the marginal steady state probability to find n users at station M. Then, according to the product form (5.18),

$$
\begin{aligned}
p_M(n; K) &= \sum_{\mathbf{n} \in E(M-1, K-n)} p_{(n_1, n_2, \dots, n_{M-1}, n)} \\
&= \sum_{\mathbf{n} \in E(M-1, K-n)} \frac{\prod_{i=1}^{M-1} x_i(n_i)\, x_M(n)}{C_M(K)} \\
&= \frac{C_{M-1}(K-n)}{C_M(K)} \cdot x_M(n). \quad (5.22)
\end{aligned}
$$

The mean number $\bar{N}_M(K)$ of customers in station M is now immediately obtained as[7]

$$\bar{N}_M(K) = \sum_{n=0}^{K} p_M(n; K) \cdot n.$$

Hence,

$$\bar{N}_M(K) = \frac{1}{C_M(K)} \sum_{n=0}^{K} C_{M-1}(K-n)\, x_M(n) \cdot n. \quad (5.23)$$

---

[7]We intentionally indicate here and in the following in each term the total number of customers present in the network.

It may be worthwile to note that this expression again takes the form of a convolution: Set $\mathbf{z}_M = (0, x_M(1), 2x_M(2), \ldots, K x_M(K))$; then

$$\bar{N}_M(K) = \frac{(\mathbf{C}_{M-1} * \mathbf{z}_M)(K)}{C_M(K)}.$$

In equilibrium, the mean throughput rate $\bar{S}_M(K)$ through station M in a GN network with K customers equals the mean arrival rate $\lambda_M(K)$ (as well as the mean departure rate $\delta_M(K)$). It is given as

$$\bar{S}_M(K) = \sum_{n=1}^{K} p_M(n; K) \, \mu_m(n).$$

From this expression, by inserting (5.21) and exploiting (5.19), we obtain

$$\bar{S}_M(K) = \frac{C_M(K-1)}{C_M(K)} \cdot y_M. \tag{5.24}$$

We proceed to calculate the mean time $\bar{T}_M(K)$ a user spends in a station i (mean system time, or mean sojourn time). According to Little's result we have $\bar{T}_M(K) = \frac{1}{\lambda_M} \bar{N}_M(K) = \frac{\bar{N}_M(K)}{\bar{S}_M(K)}$, and so the above results yield

$$\bar{T}_M(K) = \frac{\sum_{n=1}^{K} C_{M-1}(K-n) \, x_M(n) \, n}{C_M(K-1) \, y_M}. \tag{5.25}$$

The mean number $\bar{N}_i^Q(K, s_i)$ of customers waiting in the queue at some station i that has s_i exponential servers is given as $\sum_{n_i=s_i}^{K} p_i(n_i; K) \, (n_i - s_i)$. Thus, by (5.21),

$$\bar{N}_M^Q(K, s_i) = \frac{\sum_{n=s_M}^{K} C_{M-1}(K-n) \, x_M(n) \, (n - s_M)}{C_M(K)}. \tag{5.26}$$

The Principle of Mean Value Analysis

In case that each network station either is a single server station or an infinite server station, an even easier way can be pursued, avoiding the explicit computation of the values (5.21). In fact, it is possible to obtain all mean values by some simple iteration process. For stations with more than 1 and less than K servers, but, one still has to rely on (5.21) and related expressions. The approach in question, in its general form, is called *mean value analysis* (MVA) and has been suggested by Reiser and Lavenberg in 1980 [73].

For a GN network the mean visiting numbers satisfy the equations

$$v_i = \sum_{j=1}^{M} v_j \, q_{ji}, \quad 1 \le i \le M. \tag{5.27}$$

These equations, clearly, do not possess a unique solution, as has been the case for an open network. So, neither we can determine the exact values for the visiting numbers v_i and the mean arrival rates λ_i, nor we can compute other mean values by imitating the previous approach. In order to achieve yet similar results, we proceed by turning a closed network into an open one without changing any of the performance criteria. The idea is the following: Add another fictitious node 0 to the network graph between two nodes i_0 and j_0 that are connected by an edge (possibly $i_0 = j_0$), where $i_0, j_0 \in \{1, \ldots, M\}$ (see figure 5.2).

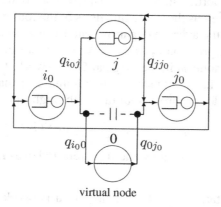

virtual node

Figure 5.2. Modified Network

Any customer, who is routed to node j_0 after service completion at node i_0 is now assumed to depart from the network, and to be immediately replaced by another new customer who enters the network at node j_0. This way, the number K of customers in the network is preserved all the time, and all network parameters remain exactly the same. The construction allows to speak of performance items like network delay T (i.e. the total time a customer spends in the network), or throughput S through the network. In particular, we shall be able to calculate the mean values $\bar{T}$, $\bar{S}$ and $\bar{N} = \sum_{i=1}^{M} \bar{N}_i$,[8] and to determine the mean number v_i of visits that a customer makes at station i for $i \in \{1, \ldots, M\}$.

---

[8]These values depend on the number K of customers in the network, and we shall indicate this dependency in the following.

First observe that any "newly arriving" customer from "outside the network" (i.e. from node 0) visits node i_0 exactly k times with probability given by $q_{i_0 j_0}(1 - q_{i_0 j_0})^{k-1}$, implying that the mean number v_{i_0} of visits at node i_0 for any customer in this open network attains the value

$$v_{i_0} = \sum_{k=1}^{\infty} k\, q_{i_0 j_0}(1 - q_{i_0 j_0})^{k-1} = \frac{1}{q_{i_0 j_0}}.$$

This determines all other visiting numbers v_i according to equation (5.27):

$$v_i = \sum_{\substack{j=1 \\ j \neq i_0}}^{M} v_j\, q_{ji} + \frac{q_{i_0 i}}{q_{i_0 j_0}}.$$

Equations (5.17) and (5.27) show that the vectors $\lambda = (\lambda_1, \ldots, \lambda_M)$ and $\mathbf{v} = (v_1, \ldots, v_M)$ are proportional, $\lambda = \gamma\,\mathbf{v}$. The λ_i (and so the constant γ) depend on the number K, whereas the v_i are functions of the routing probabilities only. Since in equilibrium the average departure rate from node i_0 equals its average arrival rate λ_{i_0}, the expression $\lambda_{i_0} q_{i_0 j_0} = \lambda_{i_0}/v_{i_0}$ represents the mean transfer rate from node i_0 to node j_0 in the original GN network. Consequently, in our artificial open network, the constant γ is nothing else than the total average input rate from outside (or throughput rate $\bar{S}$ through) the network:

$$\lambda_{i_0} q_{i_0 j_0} = \lambda_{i_0}/v_{i_0} = \gamma = \bar{S}(K).$$

The visit numbers $\lambda_i/\gamma = v_i$ are sometimes referred to as **relative throughput rates**.

Let again, for any station $i \in \{1, \ldots, M\}$ in a network with K customers, denote by $\bar{N}_i(K, s_i)$ the mean number of customers in i, and by $\bar{T}_i(K, s_i)$ the mean sojourn time in station i if there are s_i servers at that station. We obtain the following relations:

$$\bar{N}_i(K, s_i) = \lambda_i \bar{T}_i(s_i) \quad \text{(Little's rule)},$$

$$K = \sum_{i=1}^{M} \bar{N}_i(K, s_i) = \bar{S}(K) \sum_{i=1}^{M} v_i \bar{T}_i(K, s_i),$$

$$\bar{T}(K) = \sum_{i=1}^{M} v_i \bar{T}_i(K, s_i) = \sum_{i=1}^{M} \frac{\lambda_i}{\bar{S}(K)} \bar{T}_i(K, s_i).$$

The last equation confirms $K = \bar{S}(K)\bar{T}(K)$ (Little's rule). Notice, that the mean system times $\bar{T}_i(K, s_i)$ cannot be calculated as sojourn times of isolated independent $M/M/s_i$ stations as in case of a Jackson network, since the numbers $N_i(K, s_i)$ are now dependent upon each other due to the second of the above equations. Accordingly, we have to find another way to compute the $\bar{T}_i(K, s_i)$ in order to solve the equations for all other unknowns.

Let us call a customer, who has completed service at some station j and is about to enter station i (but not yet there), to be **a customer in transit** to i. Assume that such a customer "sees" $A_i(K, s_i)$ customers in total, and $A_i^Q(K, s_i)$ customers waiting in the queue at station i immediately before his entrance there. Clearly, $A_i(K, s_i)$ and $A_i^Q(K, s_i)$ are random numbers. Let $\bar{A}_i(K, s_i)$ and $\bar{A}_i^Q(K, s_i)$ denote their respective expectations. With $\bar{A}_i(K, s_i)$ the mean system times $\bar{T}_i(K, s_i)$ for the cases $s_i = 1$ and $s_i \geq K$ are given as

$$
\bar{T}_i(K, s_i) = \begin{cases} \frac{1}{\mu_i} + \frac{1}{\mu_i} \bar{A}_i(K, s_i) & \text{if} \quad s_i = 1 \\ \frac{1}{\mu_i} & \text{if} \quad s_i \geq K \end{cases} \tag{5.28}
$$

(remember, that service times are exponentially distributed, and that $1/\mu_i$ is the mean of the service time at station i). We shall show later that the corresponding value for the case $1 < s_i < K$ reads

$$
\bar{T}_i(K, s_i) = \frac{1}{\mu_i}\left(1 + \frac{1}{s_i}\left[\bar{A}_i^Q(K, s_i) + b_i(K-1)\right]\right), \quad 1 < s_i < K,
$$

where $b_i(K-1)$ is the probability for the event that, in a closed network of same type with $K-1$ customers, all of the s_i servers at station i are occupied. The task here is to compute the mean values $\bar{A}_i^Q(K, s_i)$ as well as the probabilities $b_i(K-1)$ for $i \in \{1, \ldots, M\}$.

In order to determine the $\bar{A}_i(K, s_i)$ in (5.28) we mention an important general feature of product form networks that is called the **arrival property** in case of open networks, and the **random observer property** in case of closed ones.[9] Here we confine ourself to the case of a single class GN network, but it should be clear from the proof below that this property also holds for multi-class (open or closed) PF networks.

Theorem 5.13 (Random Observer Property) *Let $a_i(\mathbf{n}-\mathbf{e}_i)$ denote the probability for the event that a customer in transit to i "sees" the state disposition $\mathbf{n} - \mathbf{e}_i = (n_1, \ldots, n_i - 1, \ldots, n_M)$ immediately before his arrival at station i. If $\mathcal{N}$ describes a closed GN network with, say, K customers, then this probability $a_i(\mathbf{n} - \mathbf{e}_i)$ is the same as the steady state probability for state $\mathbf{n} - \mathbf{e}_i$ for a network of same type with one customer less.*

---

[9]For a single queueing station this is comparable with the PASTA property (Poisson arrivals see time averages).

Proof: We denote by $\eta_i(\mathbf{n})$ the mean number of customers in transit to i per unit time who "see" the state disposition $\mathbf{n}$. Obviously, we have

$$\eta_i(\mathbf{n} - \mathbf{e}_i) = \sum_{j=1}^{M} p_{\mathbf{n} - \mathbf{e}_i + \mathbf{e}_j} \, g_{\mathbf{n} - \mathbf{e}_i + \mathbf{e}_j \, \mathbf{n}}. \qquad (5.29)$$

The probability $a_i(\mathbf{n} - \mathbf{e}_i)$ can be expressed as the relative portion of the rate $\eta_i(\mathbf{n} - \mathbf{e}_i)$ compared with the sum over all rates $\eta_i(\mathbf{m})$:

$$a_i(\mathbf{n} - \mathbf{e}_i) = \frac{\eta_i(\mathbf{n} - \mathbf{e}_i)}{\sum_{\mathbf{m} \in E(M,K)} \eta_i(\mathbf{m})}.$$

Observing $g_{\mathbf{n} - \mathbf{e}_i + \mathbf{e}_j \, \mathbf{n}} = \mu_j(n_j + 1)q_{ji}$, as well as the local balance equations

$$p_j(n_j + 1) \, \mu_j(n_j + 1) = p_j(n_j) \, \lambda_j,$$

and exploiting the product form (5.14), we obtain from (5.29), that

$$\eta_i(\mathbf{n} - \mathbf{e}_i) = p_{\mathbf{n} - \mathbf{e}_i} \sum_{j=1}^{M} \lambda_j \, q_{ji} = p_{\mathbf{n} - \mathbf{e}_i} \, \lambda_i,$$

and likewise $\eta_i(\mathbf{m}) = p_{\mathbf{m}} \lambda_i$ for any $\mathbf{m} \in E(M, K)$, $i \in \{1, \dots, M\}$. This proves $a_{\mathbf{n} - \mathbf{e}_i} = p_{\mathbf{n} - \mathbf{e}_i}$.
□

The random observer property enables us to determine the values $\bar{A}_i(K, s_i)$ and $\bar{A}_i^Q(K, s_i)$ as

$$\bar{A}_i(K, s_i) = \bar{N}_i(K - 1, s_i), \quad \bar{A}_i^Q(K, s_i) = \bar{N}_i^Q(K - 1, s_i)$$

for $i \in \{1, \dots, M\}$, where we indicate by s_i the number of servers, and by K or $K - 1$ the number of customers in the network. According to these results the station specific mean system times in a GN network with K customers are given by[10]

$$\bar{T}_i(K, s_i) = \begin{cases} \frac{1}{\mu_i} + \frac{1}{\mu_i} \bar{N}_i(K - 1, s_i) & \text{if } s_i = 1 \\ \frac{1}{\mu_i} & \text{if } s_i \geq K \\ \frac{1}{\mu_i}\left(1 + \frac{1}{s_i}\left[\bar{N}_i^Q(K - 1, s_i) + b_i(K - 1)\right]\right) & \text{if } 1 < s_i < K \end{cases}$$

$$(5.30)$$

---

[10]The third line expression for $1 < s_i < K$ will be derived later.

For a network that is built up by only single server or infinite server stations we arrive at a system of recursion equations for the unknowns $\bar{T}_i(K, s_i)$, $\bar{N}_i(K, s_i)$, and $\bar{S}(K)$, viz.

$$\bar{T}_i(K, s_i) = \begin{cases} \frac{1}{\mu_i} + \frac{1}{\mu_i} \bar{N}_i(K-1, s_i) & \text{if } s_i = 1 \\ \frac{1}{\mu_i} & \text{if } s_i \geq K \end{cases},$$

$$\bar{S}(K) = \frac{K}{\sum_{i=1}^{M} v_i \bar{T}_i(K, s_i)},$$

$$\bar{N}_i(K, s_i) = v_i \bar{S}(K) \bar{T}_i(K, s_i). \tag{5.31}$$

In essence these expressions explain what is usually meant with "mean value analysis" for Gordon-Newell networks: It is a simple iteration process that starts with $\bar{N}_i(0, s_i) = 0$ and $\bar{T}_i(1, s_i) = 1/\mu_i$ and requires to compute successively, for $1 \leq k \leq K$, the values $\bar{T}_i(k, s_i)$, $\bar{S}(K)$, and $\bar{N}_i(K, s_i)$ according to (5.31).

The computational overhead is fairly small, and can even be further reduced if approximate results are tolerated. The quantities $\bar{A}_i(K, s_i) = \bar{N}_i(K-1, s_i)$ may roughly be estimated as

$$\bar{A}_i(K, s_i) \approx \frac{K-1}{K} \bar{N}_i(K, s_i),$$

a relation that is exact for $K = 1$, and tends, for increasing K, asymptotically to an exact equation. It even provides, in many practical cases, good results for intermediate values of K. Inserted in (5.31) we obtain

$$\bar{T}_i(K, s_i) = \begin{cases} \frac{1}{\mu_i}\left(1 + \frac{v_i(K-1)\bar{T}_i(K,s_i)}{\sum_{\ell=1}^{M} v_\ell \bar{T}_\ell(K,s_\ell)}\right) & \text{if } s_i = 1 \\ \frac{1}{\mu_i} & \text{if } s_i \geq K \end{cases}.$$

So we see that, if one accepts approximate results, the overhead for the computation of the $T_i(K, s_i)$ can drastically be reduced, and limited to the solution of some fixed-point equations. It may also be the case that the total mean throughput rate $\bar{S} = \gamma$ is given in a concrete situation, meaning that we can measure somehow the average transfer rate between two connected network nodes i_0 and j_0. Then the relative throughput rates $\lambda_i/v_i = \gamma$ and, consequently, the exact arrival rates λ_i are obtained immediately, providing the $\bar{T}_i(K, s_i)$ from direct recursion:

$$\bar{T}_i(K, s_i) = \begin{cases} \frac{1}{\mu_i} + \frac{\lambda_i}{\mu_i} \bar{T}_i(K-1, s_i) & \text{if } s_i = 1 \\ \frac{1}{\mu_i} & \text{if } s_i \geq K \end{cases}.$$

We refer the reader to the literature for more detailed descriptions of the principles of mean value analysis. A practice oriented treatment, for instance, is given in the book of Bolch et al. [15], where several examples and algorithms are presented.

Example (Central Server Systems).
A closed network in which all customers are routed through some particular station before they can visit other network nodes is called a **central server system**. Examples for real configurations that may be modelled this way are computer multiprogramming systems (with a fixed degree of multiprogramming), multiprocessor systems connected with external memory modules, or a combination of independently working machines together with one single repair unit that is visited whenever one of the machines fails. The latter configuration is representative for many related ones and is known as the *machine-repairman model* (already encountered in section 6). Common to all is the possibility to model the system as a closed queueing network of the above mentioned type.

Consider, for instance, a multiprocessor system in which each processor or CPU is connected with a bank of memory modules. As soon as a processor needs some data (e.g. instructions) from a memory module it sends a request to the bank and stops working until the request is satisfied. The memory modules have buffers into which requests are arranged according to the first-come first-served (FCFS) order. A request is "served" by sending back the data to the requesting processor. Such a system can be modelled as a closed network in which the multiprocessor system represents one single infinite server (IS) station, and the, say, $M - 1$ memory modules are single server queueing stations (see figure 5.3).

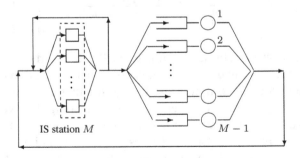

Figure 5.3.   Central Server Model

The number K of processors usually is much higher (e.g. $2^8 = 256$) than the number of memory modules. Since a processor is assumed to wait (remains in idle state) when a request to the memory bank has been sent, these requests are to be interpreted as the "users" of the single server stations numbered $1, \ldots, M - 1$, whereas intermediately executed job partitions represent the "users" in the central IS station. In many cases it is well justified to assume the time to satisfy a request being exponentially distributed. We shall see below that — with respect to an expected product form solution — there is no reason to restrict possible choices of service time distributions at the central IS server to negative exponential distributions (section 3). Nevertheless, in order to give a simple formulation, we confine ourselves to the case of an IS station with identical exponential servers with rate μ_M. As a consequence, the multiprocessor system can be modelled as a closed PF (Gordon-Newell) network.

After partially executing a job a processor may execute another partition or send a request to the memory bank. In practice, the memory bank is needed only for, say, κ times in the average. Let κ be identical for all processors. We summarize the assumptions as follows:

- Service times at the memory modules $i = 1, \ldots, M - 1$ are exponentially distributed with parameters μ_i.

- Processor execution times are exponentially distributed with mean $1/\mu_M$.

- Each processor references memory module i in the long run with probability q_{Mi}, where $\sum_{i=1}^{M-1} q_{Mi} = 1$ (central server condition).

- After κ execution times a job is finished, and starts anew (as another job) immediately after at the same processor.

Turning the closed network into an open one by applying a similar construction as mentioned in this section, the interpretation of restarts after job finishing finds its adequate portrayal in the model.

Let a virtual node (the "network exterior") be inserted between the routing edge from M to M, such that the mean number of visits to node M attains the value $v_M = 1/q_{MM} = \kappa$ (figure 5.4).

Then the routing probabilities satisfy the relations

$$q_{M0} + \sum_{i=1}^{M-1} q_{Mi} = 1,$$

$$q_{iM} = 1 \qquad \text{for all } i \in \{1, \ldots, M-1\},$$

$$q_{ij} = 0 \qquad \text{for all } i, j \in \{1, \ldots, M-1\}$$

$$q_{M0} = q_{M0} = q_{MM} = \frac{1}{v_m} = \frac{1}{\kappa}.$$

What are the performance measures to be computed? It is likely that one is interested, in the first line, in the mean time $\bar{T}$ that is required to completely

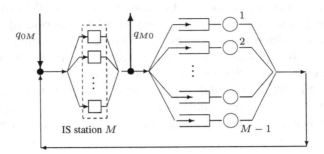

Figure 5.4.   Modified Central Server Model

execute a job, the average delay $\bar{T}_i$ at memory module i per request, the total mean throughput $\bar{S}$ of jobs through the multiprocessor system, and the average number $\bar{N}_i$ of requests waiting or being treated at some memory module i.

These quantities are easily obtained according to the mean value analysis principle: Exploiting equations (5.30), merely the corresponding iteration process has to be performed, using $\bar{N}_i(0, s_i) = 0$ as the starting value.

The General Case $1 < s_i < K$

We close this section by turning back to the general case of GN networks that contain multiple server stations with $s_i < K$. Relying on the results from convolution analysis, we can express the quantity $\bar{T}(K; s_i)$ for a customer's mean sojourn time at some station $i \in \{1, \ldots, M\}$ as a function of the quantities $\bar{N}_i^Q(K-1, s_i)$ and $b_i(K-1)$ as follows. Let a renumbering be performed such that our station under consideration has index M. According to $x_M(0) = 1$ and $x_M(n)/y_M = x_M(n-1)/\mu_M(n)$, the expression (5.25) can be rewritten as

$$
\begin{aligned}
\bar{T}_M(K) \;=\;& \sum_{n=1}^{K} \frac{C_{M-1}(K-1-[n-1])\, x_M(n-1)\, n}{C_M(K-1)\, \mu_M(n)} \\[2mm]
=\;& \sum_{n=1}^{s_M} \frac{C_{M-1}(K-1-[n-1])\, x_M(n-1)\, n}{C_M(K-1)\, n\, \mu_M} \\[2mm]
& +\; \sum_{n=s_M+1}^{K} \frac{C_{M-1}(K-1-[n-1])\, x_M(n-1)\, n}{C_M(K-1)\, s_M\, \mu_M}
\end{aligned}
$$

$$= \sum_{n=0}^{K-1} \frac{C_{M-1}(K-1-n)\,x_M(n)}{C_M(K-1)\,\mu_M}$$

$$+ \frac{1}{s_M \mu_M} \sum_{n=s_M+1}^{K} \frac{C_{M-1}(K-1-n)\,x_M(n-1)\,(n-s_M)}{C_M(K-1)}.$$

Exploiting $\sum_{n=0}^{K-1} C_{M-1}(K-1-n)x_M(n) = (\mathbf{C}_{M-1} * \mathbf{x}_M)(K-1) = C_M(K-1)$ in the first term, and setting $n - s_M = (n-1) - s_M + 1$ in the second one, we obtain from (5.21) and (5.26)

$$\bar{T}_M(K) = \frac{1}{\mu_M} + \frac{1}{\mu_M\,s_M} \left(\sum_{n=s_M}^{K-1} (n-s_M)\frac{C_{M-1}(K-1-n)\,x_M(n)}{C_M(K-1)} \right.$$

$$\left. + \sum_{n=s_M}^{K-1} \frac{C_{M-1}(K-1-n)\,x_M(n)}{C_M(K-1)} \right)$$

$$= \frac{1}{\mu_M} + \frac{1}{\mu_M\,s_M} \left(\bar{N}_M^Q(K-1, s_M) + \sum_{n=s_M}^{K-1} p_M(n; K-1) \right).$$

The sum $\sum_{n=s_M}^{K-1} p_M(n; K-1)$ represents the probability for the event that, in a network with $K - 1$ customers, at least as many customers are present at station M as there are servers, which is nothing else than the probability $b_M(K-1)$ for the event that all servers s_M are occupied. Hence, replacing the index M by an arbitrary station index $i \in \{1, \ldots, M\}$, we have

$$\bar{T}_i(K) = \frac{1}{\mu_i} \left(1 + \frac{1}{s_i} \left[\bar{N}_i^Q(K-1, s_i) + b_i(K-1) \right] \right). \tag{5.32}$$

The remaining mean values are given by equations (5.23), (5.24), and (5.26). In principle, all performance measures of a Gordon-Newell network with K customers can be calculated from the corresponding expressions for a network with one customer less, as is obvious from equations (5.30) and (5.32).

3. Symmetric Service Disciplines

Consider a queueing network with several chains and R customer classes, and remember that the pair (r, c) of class and chain identifiers defines the *category* of a customer. Let $\mu_{jrc}^{(i)}(\mathbf{n}_i)$ denote the mean service rate for a category (r, c) customer at position j in station i, when the latter is in state $\mathbf{n}_i$, and denote by $\pi_{jrc}^{(i)}(\mathbf{n}_i)$ the probability for the event that a category (r, c) customer in transit to i is going to enter this very position j when immediately before his entrance the state is $\mathbf{n}_i$. Further, let $\mu^{(i)}(\mathbf{n}_i)$ be the total mean service rate at station i in that state. Then

$$\frac{\mu_{jrc}^{(i)}(\mathbf{n}_i)}{\mu^{(i)}(\mathbf{n}_i)} =: \varphi_{jrc}^{(i)}(\mathbf{n}_i)$$

represents the fraction of the service rate that category (r, c) customers in position j produce at station i in state $\mathbf{n}_i$.

Definition 5.14 The service discipline at a station i is called a **symmetric service discipline**, and the station is said to be in **station balance**, if

$$\varphi_{jrc}^{(i)}(\mathbf{n}_i + \mathbf{e}_{rc}) = \pi_{jrc}^{(i)}(\mathbf{n}_i), \tag{5.33}$$

that is, if the service rate $\mu_{jrc}^{(i)}(\mathbf{n}_i + \mathbf{e}_{rc})$ is proportional to the probability $\pi_{jrc}^{(i)}(\mathbf{n}_i)$.[11]

The main difference between station balance and local balance lies in the fact that station balance, in comparing rates, links together the position of a customer who completes service and the position that an arriving customer is about to occupy, whereas local balance just relates arrival and departure rates for customers of same type. Two conclusions are immediately to be drawn from this fact:

- Station balance implies local balance, but not vice versa.

- A non-exponential service discipline can only be symmetric if any arriving customer receives service immediately, i.e. as soon as he enters the system.

We now give some examples for symmetric disciplines. Thereby, in order to illustrate the relationships between the *arrangement probability upon arrival* and the *fraction of service rate at some position*, we confine ourselves to the case that only one class and only one chain exists, such that $\varphi_{jrc}^{(i)}(\mathbf{n}_i + \mathbf{e}_{rc}) =: \varphi_j^{(i)}(n_i + 1)$, and the condition for symmetry reads

$$\varphi_j^{(i)}(n_i + 1) = \pi_j^{(i)}(n_i).$$

The reader should realize that this simplification is unimportant for the examples given below, and that symmetry also holds in these cases when there are several chains and several classes.

---

[11] Again, the vector $\mathbf{e}_{rc}$ is defined as to contain a 1 at the entry of $\mathbf{n}_i$ that belongs to the category (r, c), and zeros anywhere else.

1. **Processor sharing (PS) discipline.** This is the limiting case for $\tau \to 0$ of
   a Round Robin discipline that provides service to each customer in form of
   time slices of duration τ. Positions in the queue remain undefined and can
   be assumed to be equal. For the fraction of service in state $n_i + 1$ (i.e. when
   there are $n_i + 1$ customers in station i) we have, for any $j \in \{1, \ldots, n_i + 1\}$,

$$\varphi_j^{(i)}(n_i + 1) = \frac{1}{n_i + 1}.$$

If the state is n_i immediately before an arrival, then the newly arriving customer can be arranged in any of $n_i + 1$ positions with same probability. So,
$\pi_j^{(i)}(n_i) = \frac{1}{n_i+1} = \varphi_j^{(i)}(n_i + 1)$, and the discipline proves to be symmetric.

2. **Infinite servers (IS) discipline.** As in case of processor sharing the position of a customer doesn't play any role. The fraction $\varphi_j^{(i)}(n_i + 1)$ of the
   service rate that an $(n_i + 1)^{th}$ customer receives is always the $(n_i + 1)^{th}$
   part of the total service rate in this state, viz. $\varphi_j^{(i)}(n_i + 1) = 1/(n_i + 1)$. The
   position where to be inserted is not important for an arriving customer, and
   may be seen to be equal for each position among, or in front of, or behind,
   the n_i existing customers in the station. Thus, the probability $\pi_j^{(i)}(n_i)$ is
   the same for all j, and $\pi_j^{(i)}(n_i) = 1/(n_i + 1) = \varphi_j^{(i)}(n_i + 1)$, showing the
   symmetry also in this case.

3. **Last-come first-served preemptive-resume (LCFS-PR) discipline.** In this
   discipline any newly arriving customer ousts the one in service from his
   place. Let the position of the customer in service be 1. Then the fraction
   $\varphi_1^{(i)}(n_i + 1)$ of service that the arriving customer receives, is one since all
   other customers are not served during state $n_i + 1$, i.e. $\varphi_1^{(i)}(n_i + 1) = 1$.
   On the other side, the probability $\pi_1^{(i)}(n_i)$ for the event that an arriving customer is arranged in position 1 at station i (when station i was in state n_i immediately before his arrival) is one, too. Therefore, $\varphi_1^{(i)}(n_i + 1) = \pi_1^{(i)}(n_i)$.

A special role plays the **first-come first-served (FCFS) discipline.** As is easily seen, this discipline is not symmetric, since an arriving customer is always
added to the queue at its end whereas service is provided only to the customer
at its front (first) position. If customers deserve service from different service time distributions then (5.33) cannot be satisfied for all. There is, but,
one exception: If service times are chosen from the same exponential distribution for all customers, then positions and customers are indistinguishable,
and the actual service completion rate at any time, also at an arrival instant,

remains the same due to the memoryless property of the exponential distribution. That means that the rate / fraction equation (5.33) holds. Consequently, an FCFS station providing exponential service with the same intensity to all its customers attains station balance.

A network station that provides service according to one of the above disciplines is called a PS station, IS station, LCFS-PR station, or FCFS exponential station, respectively.

We are now in the position to conclude, that a multiple chain/multiple class queueing network that is fed by Poisson arrival streams (if open or mixed), and is built up by stations of types PS, LCFS-PR, IS, or FCFS exponential, attains station balance and, consequently, local balance and product form.

This result has first been proven by Baskett, Chandy, Muntz, and Palacios in 1977 [8], and is well known as the BCMP theorem. The authors introduced a numbering for the four types of service disciplines that has been adopted by most experts in the field. It runs as follows:

- **Type 1 service**: The service discipline is FCFS, and all customers have the same negative-exponential service time distribution. The service rate may depend on the number of customers at the station (this is the case when there are more than one servers available).

- **Type 2 service**: The service discipline is PS, there is a single server at the station, and each class of customer may have a distinct service time distribution. The service time distributions have rational Laplace transform.

- **Type 3 service**: The number of servers at the station is greater than or equal to the maximum number of customers that may visit this station (IS discipline). Each class of customer may have a distinct service time distribution. The service time distributions have rational Laplace transform.

- **Type 4 service**: There is a single server at the station, the service discipline is LCFS-PR, and each class of customer may have a distinct service time distribution. The service time distributions have rational Laplace transform.

The BCMP theorem explicitly describes the factors of the product form for closed, open or mixed networks (with Poisson arrival streams). In order to present these results adequately we have to explain some details. First, two types of arrival process are distinguished: A single Poisson arrival stream whose intensity γ may be a function of the state dependent total number $K(\mathbf{n})$ of customers in the network, or several chain specific Poisson streams with intensities γ_c that in turn depend on the numbers $K_c(\mathbf{n})$ of customers in the respective chains ($1 \leq c \leq V$, V the total number of chains). Second, the state descriptions are type specific as follows:

1. For type 1 service stations (exponential service, undistinguishable customers) the queue specific states are represented by the vectors

$$\mathbf{n}_i = (r_{i1}, \ldots, r_{in_i}),$$

where n_i is the total number of customers present at station i, and r_{ij} is the class of the customer at position j in the queue. Positions are counted beginning from the "server position" 1 up to the end of the queue n_i. The need for discriminating between classes will become clear below when we specify $f_i(\mathbf{n}_i)$.

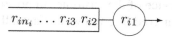

Figure 5.5. FCFS Order

2. For types 2 and 3 the service time distributions have rational Laplace transform, so they belong to the family of Cox-distributions. A Cox distribution is characterized by a sequence of exponential stages that are visited by the customer in service in compliance with routing probabilities $\alpha_{ir\ell}$. These probabilities (or their complementary values $1 - \alpha_{ir\ell}$) steer the customer to the next stage or to exit from service (see figure 5.4). Here r is the class index of the customer in service, and i the station index.

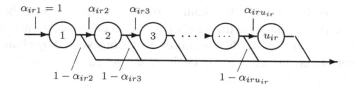

Figure 5.6. Cox Distribution

The state vector $\mathbf{n}_i$ of station i takes the form $\mathbf{n}_i = (\mathbf{s}_{i1}, \ldots, \mathbf{s}_{iR})$, where each $\mathbf{s}_{ir} = (s_{ir1}, \ldots, s_{iru_{ir}})$ is a vector of labels $s_{ir\ell}$, and

$$s_{ir\ell} = \begin{cases} \text{number of class } r \text{ customers (if any) at} \\ \text{station } i, \text{ who are in stage } \ell \text{ of service.} \end{cases}$$

$s_{ir\ell}$ is set to zero if there are no class r customers in station i. u_{ir} is the number of exponential stages for a class r service time distribution at i ($1 \leq r \leq R$).

3 Type 4 centers are characterized by the LCFS-PR scheduling discipline, offering service according to Cox distributed service times. Whereas in case of PS or IS disciplines (types 2 and 3) the customer position has no significance, here it is very important. The so-called LCFS order has the opposite direction of FCFS order (see figure 5.5). The state vector $\mathbf{n}_i$ reflects the classes as well as the stages of service of all the customers at their respective positions,

$$\mathbf{n}_i = \Big((r_1, \ell_1), (r_2, \ell_2), \ldots, (r_{n_i}, \ell_{n_i})\Big).$$

n_i is the total number of customers in station i in that state, r_j is the class, and ℓ_j the stage of service of the customer in position j. Position n_i is that of the customer who arrived last and who is actually in service.

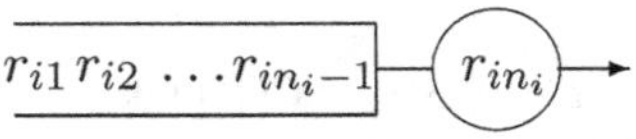

Figure 5.7.   LCFS Order

If there exist open chains in the network, then one may count the customer visits to the stations. The mean visit number to station i of a class r customer who belongs to chain c is defined as the ratio

$$v_{irc} = \frac{\lambda_{irc}}{\gamma_c(K_c(\mathbf{n}))},$$

where λ_{irc} is the mean arrival rate of category (r, c) customers at station i, and $\gamma_c(K_c(\mathbf{n}))$ is the total chain c arrival rate from outside the network, that may be dependent upon the number $K_c(\mathbf{n})$ of class c customers in the network during state $\mathbf{n}$. Let M_c be the subset of stations visited by chain c, and R_c the subset of classes occurring in chain c, and set $E_c = M_c \times R_c$. Then the λ_{irc} satisfy the traffic equations

$$\lambda_{irc} = \gamma_c(K_c(\mathbf{n})) \, q_{0;irc} + \sum_{(j,s) \in E_c} \lambda_{jsc} \, q_{jsc;irc}.$$

Consequently, the mean visit numbers v_{irc} (also called the relative throughputs) satisfy the equations

$$v_{irc} = q_{0;irc} + \sum_{(j,s) \in E_c} v_{jsc} \, q_{jsc;irc}. \tag{5.34}$$

We are now in the position to formulate the result of Baskett, Chandy, Muntz, and Palacios. Let R be the total number of customer classes, V the total number of chains, and $A_{ir\ell}$ the product of steering probabilities in a Cox distribution (compare figure 5.6), i.e., for $1 \leq i \leq M$, $1 \leq r \leq R$, and $1 \leq \ell \leq u_{ir}$,

$$A_{ir\ell} = \prod_{\nu=1}^{\ell} \alpha_{ir\nu}.$$

Theorem 5.15 (BCMP theorem) *Let an open, closed, or mixed queueing network with V chains and R customer classes contain service stations of types 1, 2, 3, or 4, only. Assume, that in case of an open or mixed network the external arrival streams are Poisson of type 1 or 2, respectively. Then, the network attains equilibrium with a product form steady state distribution*

$$p_{\mathbf{n}} = \frac{d(\mathbf{n})}{C} \prod_{i=1}^{M} f_i(\mathbf{n}_i), \tag{5.35}$$

where the $f_i(\mathbf{n}_i)$ are service type dependent state functions, and the value $d(\mathbf{n})$ is defined by

$$d(\mathbf{n}) = \begin{cases} 1 & \text{for a closed network (only 1 chain,} \\ & \text{no external arrival process)} \\ \prod_{k=0}^{K(\mathbf{n})-1} \gamma(k) & \text{for an open network (only 1 chain,} \\ & \text{external arrival process of first type)} \\ \prod_{c=1}^{V} \prod_{k=0}^{K_c(\mathbf{n})-1} \gamma_c(k) & \text{for a mixed network (several chains,} \\ & \text{external arrival processes of second type)} \end{cases}$$

The state functions $f_i(\mathbf{n}_i)$ are given by the following expressions.

Type 1 (FCFS exponential):

$$f_i(\mathbf{n}_i) = \prod_{j=1}^{n_i} \frac{v_{ir_{ij}}}{\mu_i(j)}.$$

Type 2 (PS, Cox distribution):

$$f_i(\mathbf{n}_i) = n_i! \prod_{r=1}^{R} \prod_{\ell=1}^{u_{ir}} \left(\frac{v_{ir} A_{ir\ell}}{\mu_{ir\ell}} \right)^{s_{ir\ell}} \frac{1}{s_{ir\ell}!}.$$

Type 3 (IS, Cox distribution):

$$f_i(\mathbf{n}_i) = \prod_{r=1}^{R} \prod_{\ell=1}^{u_{ir}} \left(\frac{v_{ir} A_{ir\ell}}{\mu_{ir\ell}} \right)^{s_{ir\ell}} \frac{1}{s_{ir\ell}!}.$$

Type 4 (LCFS-PR Cox distribution):

$$f_i(\mathbf{n}_i) = \prod_{j=1}^{n_i} \frac{v_{ir_{ij}} A_{ir_{ij}\ell_j}}{\mu_{ir_{ij}\ell_j}}.$$

Essentially, the BCMP theorem is a consequence from the fact that station balance implies local balance and product form. The detailed elaboration of the above mentioned concrete expressions for the factors in (5.35) can be performed by applying the symmetry relations (5.33) and the resulting local balance equations to the product of state probabilities of isolated stations, just as in case of Jackson or Gordon-Newell networks. This line of reasoning has been pursued by Baskett, Chandy, Muntz, and Palacios. We do not repeat this here, rather we refer to their original work in [8].

Notes

There is a multitude of additional results on queueing networks, including various algorithms for the exact and approximate treatment of product form (PF) networks, refined approximation methods for non-PF networks, generalizations to networks with blocking, approximation techniques for networks with priority handling, and even maximum entropy methods. To cover all these results would go far beyond of the scope of this introductory book. The elements of queueing network theory can already be found in Kleinrock's fundamental book on queueing systems (Volume I: Theory) [50], and in an early overview on exact and approximate methods for the evaluation of steady state probabilities of Markovian networks (including Jackson, Gordon-Newell, and BCMP networks) by Gelenbe and Pujolle in 1987 [36]. Also in 1987 appeared the excellent little introduction to performance analysis methods for computer communication systems by I. Mitrani [60]. The beginner is well advised to read this book first. It presents a neatly formulated and easy to understand explanation of the basic ideas behind various fundamental approaches.

A standard work on reversibility properties and their relationships to balance behaviour is that of Kelly of the year 1979 [48]. The various techniques developed there are employed also by Nelson in his recommended treatise on probability, stochastic processes, and queueing theory [61]. We further refer to the more application oriented books of Van Dijk [30], who addresses the physical background of flow balance properties, and Harrison and Patel [41] who — with respect to queueing networks — describe several applications to computer networks and computer architectures. A more recently published comprehensive treatment of queueing networks and Markov chains is that of Bolch et alii [15]. This book covers all main aspects of modern queueing network analysis, and presents up to date algorithmic methods. The reader may also find an exhaustive list of references in [15]. Finally, we refer to the excellent investigation of queueing networks with discrete time scale that has been presented in 2001 by Daduna [29]. Due to the discrete structure of most of todays communication systems this approach should attain particular attention in the future.

Exercise 5.1 Show that a stationary Markov process whose undirected state transition diagram forms a tree is reversible. Hint: Use the fact that the probability flux in one direction across a cut of the graph equals the flux in opposite direction.

Exercise 5.2 Show that any stationary birth-death process is quasi-reversible, and conclude from this fact the Theorem of Burke for $M/M/s$ queues: The departure process of an $M/M/s$ queue is Poisson with same rate as the arrival process.

Exercise 5.3 Prove that a quasi-reversible process need not be reversible. Hint: Consider an $M/M/1$ queue with mean arrival rate λ, whose state 1 is separated in two different states $1'$ and $1''$, such that $1'$ is reached from state 0 with rate $\lambda \cdot p$, and state $1''$ is reached from state 0 with rate $\lambda \cdot (1 - p)$ for $0 < p < 1$, the departure rates remaining unchanged.

Exercise 5.4 A data transmission unit works as follows. Data packages arrive at the unit according to a Poisson process with intensity λ. For each data package there is an exponential time (with parameter μ) from the beginning of transmission to the receipt of an acknowledgement. Arriving packages which find the transmission unit busy wait in a queue and are served in FCFS order. The buffer for the queue is so large that it may be assumed to be infinite. With probability p, a data package incurs a transmission error and needs to be retransmitted. The stream of data packages to retransmitted is added to the regular arrival stream.
a) Derive a model for this kind of data transmission in terms of a Jackson network.
b) Show that the combined stream of regularly arriving packages and the packages to be retransmitted is not a Poisson process.
c) Determine the mean time needed for a successful transmission of a data package in the stationary regime.

Exercise 5.5 A server in a computer pool is modelled as a queueing network with two stations. The first of these represents the CPU, the second one all output devices. Service times in both stations are distributed exponentially, with parameters μ_1 and μ_2. Jobs arrive from the pool as a Poisson process with intensity λ. After service in the CPU, a job is done with probability p. With probability $1 - p$ it needs additional service by one of the output devices.

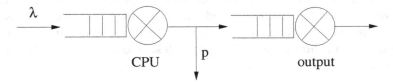

Figure 5.8. Simple model of a computer pool

Determine the mean sojourn time of a job in the server under the stationary regime. Networks consisting of some stations in series are called **tandem queues**. However, in general tandem queues the service times do not need to be exponential.

Exercise 5.6 For a **cyclic closed network** with M stations, the routing matrix Q is given by

$$Q(i,j) := \begin{cases} 1, & j = i + 1, 1 \le i < M \\ 1, & j = 1, i = M \\ 0, & \text{else} \end{cases}$$

Assume that there are K users in the network. Show that the stationary distribution is given by

$$p(n) = \frac{1}{G(K)} \cdot \frac{\mu_1^{K-n_1}}{\prod_{i=2}^{M} \mu_i^{n_i}}$$

with μ_i denoting the service rate at station i.

Exercise 5.7 An internet based company seeks to ensure constant online access, because it cannot operate without. To this aim, two servers instead of one are employed concurrently. Each of them has a **failure rate** $\lambda > 0$, meaning that their up time is exponentially distributed with parameter λ. After failure, a server is repaired with probability p. The repair time is distributed exponentially with parameter $\mu_1 > 0$. With probability $1 - p$, the server must be replaced by a new one, which requires an installation time that is distributed exponentially with parameter $\mu_2 > 0$. After the server is repaired, there is still a probability q that it must be replaced by a new one, requiring additionally the same installation time.

Derive a model for this situation in terms of a Gordon–Newell network. For the values $\lambda = 2$, $\mu_1 = 1$, $\mu_2 = 3$, $p = 3/4$, and $q = 1/3$, determine the stationary probability that both servers are down and the company cannot operate. Compare this to the stationary probability that the company cannot operate for the case that only one server is employed. Such questions are typical for **reliability theory**.

PART II

SEMI–MARKOVIAN METHODS

Chapter 6

RENEWAL THEORY

1. Renewal Processes

Be $(X_n : n \in \mathbb{N}_0)$ a sequence of independent positive random variables, and assume that $(X_n : n \in \mathbb{N})$ are identically distributed. Define the sequence $\mathcal{S} = (S_n : n \in \mathbb{N})$ by $S_1 := X_0$ and $S_{n+1} := S_n + X_n$ for all $n \in \mathbb{N}$. The random variable S_n, with $n \in \mathbb{N}$, is called the nth **renewal time**, while the time duration X_n is called the nth **renewal interval**. Further define the random variable of the number of renewals until time t by

$$N_t := \max\{n \in \mathbb{N} : S_n \leq t\}$$

for all $t \geq 0$ with the convention $\max \emptyset = 0$. Then the continuous time process $\mathcal{N} = (N_t : t \in \mathbb{R}_0^+)$ is called a **renewal process**. The random variable X_0 is called the **delay** of $\mathcal{N}$. If X_0 and X_1 have the same distribution, then $\mathcal{N}$ is called an **ordinary renewal process**.

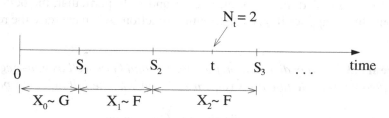

Figure 6.1. Random variables of a renewal process

We will always assume that $\mathbb{P}(X_1 = 0) = 0$ and $m := \mathbb{E}(X_1) < \infty$ is finite. The strong law of large numbers implies that $S_n/n \to m$ with probability one as $n \to \infty$. Hence $S_n < t$ cannot hold for infinitely many n and thus N_t is finite with probability one. By standard notation we will write

$$G(x) := \mathbb{P}(X_0 \leq x) \qquad \text{and} \qquad F(x) := \mathbb{P}(X_1 \leq x)$$

for all $x \in \mathbb{R}_0^+$.

Example 6.1 A light bulb has been installed at time zero. After a duration X_0, it will go out of order. We assume that it will be immediately replaced by a new light bulb at time $S_1 = X_0$. Assume that the new light bulb is of a type identical to the old one. Then the duration X_1 until it goes out of order is distributed identically to X_0. Of course, the life times of the light bulbs are independent from one another. Keeping up this rechangement policy over time, the number N_t of used light bulbs until time t forms an ordinary renewal process.

Remark 6.2 A Poisson process with intensity λ (see example 3.1) is an ordinary renewal process with $F(x) = G(x) = 1 - e^{-\lambda x}$, i.e. the renewal intervals have an exponential distribution. Thus a renewal process can be seen as a generalization of the Poisson process with respect to the distribution of the renewal intervals.

In order to derive an expression for the distribution and the expectation of N_t at any time t, we need to introduce the concept of **convolution**s of a non–negative function and a distribution function. Let F denote a distribution function on $\mathbb{R}_0^+$ and $g : \mathbb{R}_0^+ \to \mathbb{R}_0^+$ a Lebesgue–measurable function which is bounded on all finite intervals $[0, t]$ with $t \geq 0$. Then the function defined by

$$F * g(t) := \int_0^t g(t - u) \, dF(u)$$

for all $t \in \mathbb{R}$ is called the convolution of F and g. In particular, the definition of a convolution applies if g is a distribution function. As an exercise the reader can prove

Theorem 6.3 *For any distribution functions F and G as well as non–negative Lebesgue–measurable functions $(g_n : n \in \mathbb{N})$ on $\mathbb{R}_0^+$, the following properties hold:*
(1) The convolution $F * G$ is a distribution function on $\mathbb{R}_0^+$.
(2) $F * G = G * F$
(3) $F * \sum_{n=1}^{\infty} g_n = \sum_{n=1}^{\infty} F * g_n$
(4) The Dirac measure δ_0 on 0 with distribution function I_0, which is defined by $I_0(t) := 1$ for all $t \geq 0$ and $I_0(t) := 0$ otherwise, is neutral in regard to convolutions, i.e. $I_0 * G = G$ for all distribution functions G.

*(5) If the random variables X and Y are independent and distributed accord-
ing to F and G, respectively, then $\mathbb{P}(X + Y \le t) = F * G(t)$ for all $t \ge 0$.
(6) $F * (G * g) = (F * G) * g$*

Let F denote any distribution function for a real–valued random variable. Define the **convolutional power**s by $F^{*1} := F$ and recursively $F^{*n+1} := F^{*n} * F$
for all $n \in \mathbb{N}$. Because of property (4) in the above theorem, we define
$F^{*0} := I_0$ for every distribution function F.

Now denote the distribution function of the random variable X_1 (and hence
of all X_n with $n \ge 1$) and X_0 by F and G, respectively. Since the random
variables $(X_n : n \in \mathbb{N})$ are iid, part (5) of the above theorem yields for all
$n \in \mathbb{N}_0$ the relation $\mathbb{P}(N_t \ge n) = \mathbb{P}(S_n \le t) = G * F^{*n-1}(t)$ and thus we
obtain $\mathbb{P}(N_t = 0) = 1 - G(t)$ and

$$\mathbb{P}(N_t = n) = \mathbb{P}(S_n \le t) - \mathbb{P}(S_{n+1} \le t) = G * F^{*n-1}(t) - G * F^{*n}(t)$$

for $n \ge 1$. The expectation of N_t is given by

$$\mathbb{E}(N_t) = \sum_{n=1}^{\infty} \mathbb{P}(N_t \ge n) = \sum_{n=1}^{\infty} \mathbb{P}(S_n \le t) = G * \sum_{n=0}^{\infty} F^{*n}(t) \qquad (6.1)$$

for all $t \ge 0$ (for the first equality see Exercise 6.2). The rate of growth of a
renewal process is described by

Theorem 6.4 *Let $\mathcal{N} = (N_t : t \ge 0)$ denote a renewal process with renewal
intervals having mean length $m < \infty$. Then*

$$\lim_{t \to \infty} \frac{N_t}{t} = \frac{1}{m}$$

holds with probability one.

Proof: By definition of N_t (see picture below and figure 1), the inequalities
$S_{N_t} \le t \le S_{N_t+1}$ hold with probability one for all times t.

Dividing these by N_t and using the strong law of large numbers, we obtain

$$m = \lim_{n \to \infty} \frac{S_n}{n} = \lim_{t \to \infty} \frac{S_{N_t}}{N_t}$$

$$\le \lim_{t \to \infty} \frac{t}{N_t}$$

$$\le \lim_{t \to \infty} \left(\frac{S_{N_t+1}}{N_t + 1} \cdot \frac{N_t + 1}{N_t} \right) = \lim_{n \to \infty} \frac{S_{n+1}}{n + 1} \cdot \lim_{n \to \infty} \frac{n + 1}{n} = m \cdot 1$$

which proves the statement.
□

Because of this theorem, the inverse $1/m$ of the mean length of a renewal interval is called the **rate** of the renewal process. It describes the asymptotic rate at which renewals occur.

Example 6.5 Regarding a Poisson process $\mathcal{N} = (N_t : t \geq 0)$ with intensity $\lambda > 0$, it can be shown that

$$\mathbb{P}(N_t = n) = \frac{(\lambda t)^n}{n!} e^{-\lambda t} \tag{6.2}$$

for all $t \geq 0$ and $n \in \mathbb{N}_0$. The expectation of N_t is given by $\mathbb{E}(N_t) = \lambda \cdot t$.

Thus a Poisson process with intensity λ has at time t a Poisson distribution with parameter $\lambda \cdot t$. Moreover, the intensity λ is also the rate of the Poisson process, since a mean renewal interval has length $1/\lambda$.

Given an observed stream of events (e.g. job requests at a server) over some time interval of length t, we can count the number $N(t)$ of events that have occurred in this interval. If we want to model such event streams by a Poisson process, then we need to find a statistical estimator for the intensity λ. Now theorem 6.4 states that the fraction $N(t)/t$ comes close to λ for large interval lengths t. Thus a consistent statistical estimator for the intensity λ is given by $\hat{\lambda} = N(t)/t$.

Example 6.6 There is a discrete–time analogue of the Poisson process, which is called **Bernoulli process**. This is an ordinary renewal process with renewal intervals that have a geometric distribution. Given a parameter $p \in]0, 1[$, the length of the renewal intervals is distributed as $\mathbb{P}(X_1 = n) = p \cdot (1 - p)^{n-1}$ for $n \in \mathbb{N}$.

2.   Renewal Function and Renewal Equations

The function defined by $R(t) := \sum_{n=1}^{\infty} F^{*n}(t)$ for all $t \geq 0$ is called the **renewal function** of the process $\mathcal{N}$. The renewal function will play a central role in renewal theory. First we need to show that it remains finite:

Theorem 6.7 *If $F(0) < 1$, then $R(t) = \sum_{n=1}^{\infty} F^{*n}(t) < \infty$ for all $t \geq 0$.*

Proof: Since $F(0) < 1$ and F is continuous to the right, there is a number $\alpha > 0$ such that $F(\alpha) < 1$. Fix any $t \geq 0$ and choose $k \in \mathbb{N}$ such that $k \cdot \alpha > t$. Then $F^{*k}(t) \leq 1 - (1 - F(\alpha))^k =: 1 - \beta$ with $0 < \beta < 1$. Thence we obtain the bound $F^{*mk}(t) \leq (1-\beta)^m$ for any $m \in \mathbb{N}$. Since $F(0-) = 0$, we can use $F^{*n}(t) \geq F^{*h}(t)$ for all $n < h \in \mathbb{N}$. Putting these bounds together, we obtain

$$R(t) = \sum_{n=1}^{\infty} F^{*n}(t) \leq k \cdot \sum_{m=0}^{\infty} F^{*mk}(t) \leq k \cdot \sum_{m=0}^{\infty} (1-\beta)^m = \frac{k}{\beta} < \infty$$

since $\beta > 0$.
□

Theorem 6.8 *An ordinary renewal process is uniquely determined by its renewal function.*

Proof: First we take the Laplace–Stieltjes transform (LST, see appendix 3) on both sides of the equation $R(t) = \sum_{n=1}^{\infty} F^{*n}(t)$. This yields

$$\tilde{R}(s) = \sum_{n=1}^{\infty} \widetilde{F^{*n}}(s) = \tilde{F}(s) \cdot \sum_{n=0}^{\infty} (\tilde{F}(s))^n = \frac{\tilde{F}(s)}{1 - \tilde{F}(s)} \qquad (6.3)$$

for $s > 0$, or

$$\tilde{F}(s) = \frac{\tilde{R}(s)}{1 + \tilde{R}(s)}$$

and thus determines the LST $\tilde{F}(s)$ of F uniquely in terms of $\tilde{R}(s)$. Now uniqueness of the LST yields the statement.
□

For an ordinary renewal process we can derive an implicit integral equation for the renewal function, which is known as a renewal equation. Note that for an ordinary renewal process $\mathbb{E}(N_t) = R(t)$ for all times t (see (6.1) with $G = F$). Hence the function R is increasing. If we condition upon the length x of the first renewal interval X_0, we obtain

$$\mathbb{E}(N_t) = \int_0^{\infty} \mathbb{E}(N_t | X_0 = x) \, dF(x)$$

Since $\mathbb{E}(N_t | X_0 = x) = 1 + R(t-x)$ for $t \geq x$ and $\mathbb{E}(N_t | X_0 = x) = 0$ for $t < x$, we can simplify this equation to

$$R(t) = \int_0^t (1 + R(t-x)) \, dF(x) = F(t) + \int_0^t R(t-x) \, dF(x)$$

for all $t \geq 0$. A **renewal equation** is the generalized form

$$g(t) = h(t) + \int_0^t g(t-x) \, dF(x), \qquad t \geq 0 \qquad (6.4)$$

where a function h on $[0, \infty[$ and a distribution function F on $[0, \infty[$ are given and the function g on $[0, \infty[$ is unknown. The solution is given in

Theorem 6.9 *The unique solution g to equation (6.4) is given by*

$$g(t) = \int_0^t h(t - x)\, dR(x) + h(t)$$

where $R(t) = \sum_{n=1}^\infty F^{*n}(t)$ denotes the renewal function for F.

Proof: Equation (6.4) can be written as $g = h + g * F$. Because of the definition $R = \sum_{n=1}^\infty F^{*n}$ we obtain

$$F * (R * h + h) = F * h + \sum_{n=1}^\infty F^{*n+1} * h = \sum_{n=1}^\infty F^{*n} * h = R * h$$

which shows that $g = R * h + h$ is indeed a solution of (6.4).

Let g' denote another solution and define the function

$$\delta := g' - R * h - h$$

Then (6.4) implies $\delta = F * \delta$ and thus $\delta = F^{*n} * \delta$ for all $n \in \mathbb{N}$. Since $R(t) < \infty$ for any fixed $t \geq 0$, we infer that $F^{*n} \to 0$ as $n \to \infty$. Hence $\delta(t) = 0$ for all $t \geq 0$, which completes the proof.
$\square$

3. Renewal Theorems

In order to present the most powerful results of renewal theory, it will be useful to introduce stopping times and Wald's lemma. Recall from (2.3) that a random variable S with values in $\mathbb{N}_0 \cup \{\infty\}$ is called a stopping time for the sequence $\mathcal{X} = (X_0 : n \in \mathbb{N}_0)$ if

$$\mathbb{P}(S \leq n | \mathcal{X}) = \mathbb{P}(S \leq n | X_0, \ldots, X_n) \tag{6.5}$$

holds for all $n \in \mathbb{N}_0$.

Lemma 6.10 *For a renewal process $\mathcal{N}$ with delay X_0 and renewal intervals $(X_n : n \in \mathbb{N})$, the random variable N_t is a stopping time for the sequence $(X_n : n \in \mathbb{N}_0)$.*

Proof: This follows from the observation that $N_t = k$ is equivalent to

$$\sum_{n=0}^{k-1} X_n \leq t < \sum_{n=0}^{k} X_n$$

which implies that the event $N_t \leq k$ depends only on $X_0, \ldots, X_k$.

□

Lemma 6.11 Wald's Lemma

Be $\mathcal{X} = (X_n : n \in \mathbb{N}_0)$ a sequence of stochastically independent positive random variables with the same expectation $\mathbb{E}(X_n) = m$ for all $n \in \mathbb{N}$. The expectations $\mathbb{E}(X_0)$ and $\mathbb{E}(X_1)$ shall be finite. Further be S a stopping time of the sequence $\mathcal{X}$ with $\mathbb{E}(S) < \infty$. Then

$$\mathbb{E}\left(\sum_{n=0}^{S} X_n\right) = \mathbb{E}(X_0) + \mathbb{E}(S) \cdot m$$

Proof: For all $n \in \mathbb{N}_0$ define the random variables $I_n := 1$ on the set $\{S \geq n\}$ and $I_n := 0$ else. Then $\sum_{n=0}^{S} X_n = \sum_{n=0}^{\infty} I_n X_n$ and hence

$$\mathbb{E}\left(\sum_{n=0}^{S} X_n\right) = \mathbb{E}\left(\sum_{n=0}^{\infty} I_n X_n\right) = \sum_{n=0}^{\infty} \mathbb{E}(I_n X_n)$$

by monotone convergence, as I_n and X_n are non–negative. S being a stopping time for $\mathcal{X}$, we obtain by definition $\mathbb{P}(S \geq 0) = 1$, and further

$$\mathbb{P}(S \geq n | \mathcal{X}) = 1 - \mathbb{P}(S \leq n - 1 | \mathcal{X}) = 1 - \mathbb{P}(S \leq n - 1 | X_0, \ldots, X_{n-1})$$

for all $n \in \mathbb{N}$. Since the X_n are independent, I_n and X_n are independent, too, which implies $\mathbb{E}(I_0 X_0) = \mathbb{E}(X_0)$ and

$$\mathbb{E}(I_n X_n) = \mathbb{E}(I_n) \cdot \mathbb{E}(X_n) = \mathbb{P}(S \geq n) \cdot m$$

for all $n \in \mathbb{N}$. Now the relation $\sum_{n=1}^{\infty} \mathbb{P}(S \geq n) = \mathbb{E}(S)$ yields

$$\mathbb{E}\left(\sum_{n=0}^{S} X_n\right) = \sum_{n=0}^{\infty} \mathbb{E}(I_n X_n) = \mathbb{E}(X_0) + \sum_{n=1}^{\infty} \mathbb{P}(S \geq n) \cdot m$$
$$= \mathbb{E}(X_0) + \mathbb{E}(S) \cdot m$$

□

Theorem 6.12 Elementary Renewal Theorem

Be N a renewal process with renewal intervals $(X_n : n \in \mathbb{N})$ and mean renewal time $\mathbb{E}(X_1) = m > 0$. Assume further that the mean delay is finite, i.e. $\mathbb{E}(X_0) < \infty$. Then for the counting function N_t the limit

$$\lim_{t \to \infty} \frac{\mathbb{E}(N_t)}{t} = \frac{1}{m}$$

holds, with the convention $1/\infty := 0$.

Proof: For every $t \geq 0$, the bound $t < \sum_{n=0}^{N_t} X_n$ holds almost surely. By Wald's lemma, this implies

$$t < \mathbb{E}\left(\sum_{n=0}^{N_t} X_n\right) = \mathbb{E}(X_0) + \mathbb{E}(N_t) \cdot m$$

and thence for $m < \infty$

$$\frac{1}{m} - \frac{\mathbb{E}(X_0)}{m \cdot t} < \frac{\mathbb{E}(N_t)}{t}$$

for all $t \geq 0$. For $\mathbb{E}(X_0) < \infty$ and $t \to \infty$, this yields the bound

$$\liminf_{t\to\infty} \frac{\mathbb{E}(N_t)}{t} \geq \frac{1}{m}$$

which trivially holds for the case $m = \infty$.

Now it remains to show that $\limsup_{t\to\infty} \mathbb{E}(N_t)/t \leq 1/m$. To this aim we consider the truncated renewal process, denoted by $\tilde{\mathcal{N}}$, with the same delay $\tilde{X}_0 = X_0$ but renewal intervals $\tilde{X}_n = \min(X_n, M)$ for all $n \in \mathbb{N}$, with M being a fixed constant. Denote further $\tilde{m} = \mathbb{E}(\tilde{X}_1)$.

Because of $\tilde{X}_n \leq M$ the bound $\sum_{n=0}^{\tilde{N}_t} \tilde{X}_n \leq t + M$ holds almost certainly for all $t \geq 0$. Taking expectations and applying Wald's lemma, we obtain

$$\mathbb{E}(X_0) + \mathbb{E}(\tilde{N}_t) \cdot \tilde{m} = \mathbb{E}\left(\sum_{n=0}^{\tilde{N}_t} \tilde{X}_n\right) \leq t + M$$

For $\mathbb{E}(X_0) < \infty$ and $t \to \infty$, this yields

$$\limsup_{t\to\infty} \frac{\mathbb{E}(\tilde{N}_t)}{t} \leq \frac{1}{\tilde{m}}$$

Since $\tilde{X}_n \leq X_n$ for all $n \in \mathbb{N}$, we know that $\tilde{N}_t \geq N_t$ for all $t \geq 0$. Thus we obtain further

$$\limsup_{t\to\infty} \frac{\mathbb{E}(N_t)}{t} \leq \frac{1}{\tilde{m}}$$

for any constant M. Now the result follows for $M \to \infty$.

$\square$

Remark 6.13 In view of theorem 6.4 one might be tempted to think that this trivially implied the statement of the above theorem 6.12. However, the following example shows that a limit with probability one in general does not imply a limit in expectation.

Let U denote a random variable which is uniformly distributed on the interval $]0, 1[$. Further define the random variables $(V_n : n \in \mathbb{N})$ by

$$V_n := \begin{cases} 0, & U > 1/n \\ n, & U \leq 1/n \end{cases}$$

Since $U > 0$ with probability one, we obtain the limit

$$V_n \to 0, \qquad n \to \infty$$

with probability one. On the other hand, the expectation for V_n is given by

$$\mathbb{E}(V_n) = n \cdot \mathbb{P}(U \leq 1/n) = n \cdot \frac{1}{n} = 1$$

for all $n \in \mathbb{N}$ and thus $\mathbb{E}(V_n) \to 1$ as $n \to \infty$.

A non–negative random variable X (and also its distribution function F) is called **lattice** if there is a positive number $d > 0$ with $\sum_{n=0}^{\infty} \mathbb{P}(X = nd) = 1$. If X is lattice, then the largest such number d is called the **period** of X (and F). The definition states that a lattice random variable X assumes only values that are multiples of its period d.

The next result is proven in Feller [35]. The proof is lengthy and technical and therefore not repeated.

Theorem 6.14 Blackwell's Theorem
Be $\mathcal{N}$ a renewal process with renewal intervals $(X_n : n \in \mathbb{N})$ and mean renewal time $\mathbb{E}(X_1) = m$. If X_1 is not lattice, then for any $s > 0$ the counting function N_t behaves asymptotically as

$$\lim_{t \to \infty} (\mathbb{E}(N_{t+s}) - \mathbb{E}(N_t)) = \frac{s}{m}$$

with the convention $1/\infty := 0$.

Blackwell's theorem suggests the following argument: Because of the identity $\mathbb{E}(N_t) = R(t)$, it states that asymptotically

$$R(t + s) - R(t) \to s \cdot \frac{1}{m} \qquad \text{as} \qquad t \to \infty$$

This means that increments of the renewal function $t \to R(t)$ tend to be linear (with coefficient $1/m$) for large t. If we let $s \to 0$, this would suggest

$$dR(t) \to \frac{1}{m} dt \qquad \text{as} \qquad t \to \infty$$

For functions g which behave nice enough and vanish at infinity (i.e. $g(t) \to 0$ as $t \to \infty$), we thus can hope to establish

$$\lim_{t \to \infty} \int_0^t g(t - x)\, dR(x) = \frac{1}{m} \int_0^\infty g(t)\, dt$$

In order to do this, we first need to define what we require as "nice behaviour" from g. Let $g : \mathbb{R}_0^+ \to \mathbb{R}$ denote a real–valued function on the time axis and define for $a > 0$ and $n \in \mathbb{N}$

$$M_n(a) := \sup\{g(x) : (n - 1)a \leq x \leq na\} \tag{6.6}$$

$$m_n(a) := \inf\{g(x) : (n - 1)a \leq x \leq na\} \tag{6.7}$$

The function g is called **directly Riemann integrable** if $\sum_{n=1}^\infty |M_n(a)|$ and $\sum_{n=1}^\infty |m_n(a)|$ are finite for some $a > 0$ (and then for all $0 < a' < a$), and

$$\lim_{a \to 0} a \sum_{n=1}^\infty M_n(a) = \lim_{a \to 0} a \sum_{n=1}^\infty m_n(a) \tag{6.8}$$

Remark 6.15 Direct Riemann integrability is somewhat stronger than usual Riemann integrability. The similarity is that upper and lower sums converge to the same limit as $a \to 0$. The difference is that this must happen uniformly for all intervals of the time axis.

In the rest of this book, we will deal with only two kinds of directly Riemann integrable functions. For these we provide the following lemma, which the reader may prove as an exercise.

Lemma 6.16 *Assume that $g(t) \geq 0$ for all $t \geq 0$. If either*
(1) g is non–increasing and Lebesgue integrable, or
(2) g is Riemann integrable and there is a function g^* with $g(t) \leq g^*(t)$ for all $t \geq 0$, such that g^* is directly Riemann integrable,
then g is directly Riemann integrable.

Now we can state the main result of renewal theory:

Theorem 6.17 Key Renewal Theorem

Assume that $m = \mathbb{E}(X_1) > 0$, where X_1 is not lattice, and let g denote a directly Riemann integrable function. Then

$$\lim_{t \to \infty} (R * g)(t) = \lim_{t \to \infty} \int_0^t g(t - x) \, dR(x) = \frac{1}{m} \int_0^\infty g(y) \, dy$$

holds with the convention $1/\infty = 0$.

Proof: Let $(x_n : n \in \mathbb{N}_0)$ with $x_0 := 0$ denote any countable partition of the time axis $\mathbb{R}_0^+$ into intervals of the form $I_n := [x_{n-1}, x_n[$. Define the indicator function of I_n by $i_n(t) := 1$ if $t \in I_n$ and $i_n(t) := 0$ otherwise. Then

$$(R * i_n)(t) = \int_0^t i_n(t - u) \, dR(u) = \int_{t-x_n}^{t-x_{n-1}} dR(u)$$
$$= R(t - x_{n-1}) - R(t - x_n)$$

for all $t > x_n$. Now Blackwell's theorem 6.14 yields

$$\lim_{t \to \infty} (R * i_n)(t) = \frac{x_n - x_{n-1}}{m}$$

for every $n \in \mathbb{N}$.

For any finite interval $[t - l, t[$ of length l, the interpretation that $R(t) = \mathbb{E}(N_t)$ for an ordinary renewal process $\mathcal{N}$ yields with $G_t(x) := \mathbb{P}(S_{N_t+1} - t \leq x)$ the bound

$$R(t + l) - R(t) = \mathbb{E}(N_{t+l} - N_t) = \int_0^l (R(l - x) + 1) \, dG_t(x)$$
$$\leq R(l) + 1 =: B(l) < \infty$$

for every $l > 0$.

For a function $h = \sum_{n=1}^\infty c_n i_n$ with maximal interval length M and coefficients bounded by $\sum_{n=1}^\infty c_n < \infty$, we obtain thus

$$\sum_{n=1}^k c_n \cdot (R * i_n)(t) \leq (R * h)(t) \leq \sum_{n=1}^k c_n \cdot (R * i_n)(t) + B(M) \cdot \sum_{n=k+1}^\infty c_n$$

for every $k \in \mathbb{N}$ and $t \geq 0$. Letting first $t \to \infty$ and then $k \to \infty$, the limit

$$\lim_{t \to \infty} (R * h)(t) = \frac{1}{m} \sum_{n=1}^\infty c_n \cdot (x_n - x_{n-1}) = \frac{1}{m} \int_0^\infty h(y) \, dy$$

is established for any such function h.

Since g is directly Riemann integrable, there is a family of functions

$$f_l := \sum_{n=1}^{\infty} m_n(l) i_n \quad \text{and} \quad h_l := \sum_{n=1}^{\infty} M_n(l) i_n$$

using the definitions (6.6) and (6.7). These functions satisfy $f_l \leq g \leq h_l$ and have the form of the function h above with interval length l. Then

$$R * f_l \leq R * g \leq R * h_l$$

for all $l > 0$, and the result follows for $l \to 0$ according to condition (6.8).
□

This proof shows that the key renewal theorem is a consequence of Blackwell's theorem. The simple case $g := 1_{[0,s[}$, i.e. $g(t) = 1$ for $0 \leq t < s$ and $g(t) = 0$ for $t \geq s$ yields

$$\lim_{t \to \infty} (\mathbb{E}(N_{t+s}) - \mathbb{E}(N_t)) = \lim_{t \to \infty} (\mathbb{E}(N_t - N_{t-s})) = \lim_{t \to \infty} (R * g)(t) = \frac{1}{m} \cdot s$$

as an application of the key renewal theorem. Hence the statements in Blackwell's and the key renewal theorem are equivalent.

Besides its central role in renewal theory, the key renewal theorem will serve mainly two purposes in the further presentation. First, it will give a foundation for the proof of the main limit theorem in Markov renewal theory (see chapter 7). Second, it yields a limit theorem for regenerative processes (see section 1) as an immediate corollary.

4. Residual Life Times and Stationary Renewal Processes

Choose any time $t \geq 0$. Denote the duration from t until the next arrival by $B_t := S_{N_t+1} - t$ and call it the **residual life time** (or the **excess life**) at t. Further we define $A_t := t - S_{N_t}$ and call A_t the **age** at t. The distribution of B_t appeared already in the proof of theorem 6.17.

Theorem 6.18 *Be $\mathcal{N}$ an ordinary renewal process with renewal intervals having distribution function F. Then*

$$\mathbb{P}(B_t \leq x) = F(t+x) - \int_0^t (1 - F(t+x-y)) \, dR(y)$$

for all $t \geq 0$. Further the limit

$$\lim_{t \to \infty} \mathbb{P}(B_t \leq x) = \frac{1}{m} \int_0^x (1 - F(y)) \, dy \tag{6.9}$$

holds if F is not lattice.

Proof: Fix any $x \geq 0$. First abbreviate $g(t) := \mathbb{P}(B_t > x)$ for all $t \geq 0$. Conditioning on X_0 yields

$$g(t) = \int_0^\infty \mathbb{P}(B_t > x | X_0 = s) \, dF(s)$$

By definition the event $\{B_t > x\}$ is equivalent to the event that there are no renewals in the interval $]t, t + x]$. This observation and the fact that the process restarts at $S_1 = X_0$ yield

$$\mathbb{P}(B_t > x | X_0 = s) = \begin{cases} g(t - s), & s \leq t \\ 0, & t < s \leq t + x \\ 1, & s > t + x \end{cases}$$

Hence we obtain the renewal equation

$$g(t) = \int_0^t g(t - s) \, dF(s) + 1 - F(t + x) \tag{6.10}$$

with solution

$$g(t) = \int_0^t (1 - F(t + x - y)) \, dR(y) + 1 - F(t + x)$$

This yields the first statement. The second one is obtained by using the key renewal theorem to equation (6.10). This is applicable by condition (1) of lemma 6.16 and leads to

$$\lim_{t \to \infty} g(t) = \frac{1}{m} \int_0^\infty (1 - F(t + x)) \, dt = \frac{1}{m} \int_x^\infty (1 - F(y)) \, dy$$

and because of $m = \int_0^\infty (1 - F(y)) \, dy$ we obtain further

$$\lim_{t \to \infty} \mathbb{P}(B_t \leq x) = 1 - \frac{1}{m} \int_x^\infty (1 - F(y)) \, dy = \frac{1}{m} \int_0^x (1 - F(y)) \, dy$$

$\square$

Remark 6.19 For $m = \infty$, equation (6.9) states that the residual life time asymptotically tends to infinity with probability one.

Because of the equality $\{A_t > x\} = \{B_{t-x} > x\}$, an immediate application of theorem 6.18 is

Corollary 6.20 *Be $\mathcal{N}$ an ordinary renewal process with renewal intervals having distribution function F. Then*

$$\mathbb{P}(A_t \leq x) = \begin{cases} F(t) - \int_0^{t-x}(1 - F(t-y))\, dR(y), & x < t \\ 1, & x \geq t \end{cases}$$

If F is not lattice, then the limit

$$\lim_{t\to\infty} \mathbb{P}(A_t \leq x) = \frac{1}{m}\int_0^x (1 - F(y))\, dy$$

holds.

Remark 6.21 The above results show that the distributions of age and residual life time asymptotically tend to be the same. For $m = \infty$ the same phenomenon as for the residual life time happens: The age asymptotically tends to infinity with probability one.

Theorem 6.22 *If F is not lattice and $\mathbb{E}(X_1^2) < \infty$, then the limit*

$$\lim_{t\to\infty} \mathbb{E}(B_t) = \frac{\mathbb{E}(X_1^2)}{2m}$$

holds.

Proof: Define the functions $g(t) = \mathbb{E}(B_t)$ and $h(t) := \mathbb{E}\left(B_t \cdot 1_{\{X_0 > t\}}\right)$ for all $t \geq 0$. Then the renewal equation

$$g(t) = h(t) + \int_0^t g(t-x)\, dF(x)$$

holds. The function h is positive, not increasing, and integrable with

$$\int_0^\infty h(t)\, dt = \int_{t=0}^\infty \int_{x=t}^\infty (x-t)\, dF(x)\, dt$$
$$= \int_{x=0}^\infty \int_{t=0}^x (x-t)\, dt\, dF(x) = \int_{x=0}^\infty \frac{x^2}{2}\, dF(x)$$
$$= \frac{\mathbb{E}(X_1^2)}{2}$$

Thus the key renewal theorem applies (due to condition (1) of lemma 6.16) and yields

$$\lim_{t\to\infty} \mathbb{E}(B_t) = \frac{1}{m}\int_0^\infty h(t)\, dt = \frac{\mathbb{E}(X_1^2)}{2m}$$

which completes the proof.

□

For a **stationary renewal process** we would postulate that the distribution of the counts in an interval $[s, s + t]$ be independent of s and thus equal the distribution of N_t. If this holds for a process $\mathcal{N}$, then we also say that $\mathcal{N}$ has **stationary increments**. This implies in particular that the distribution of the residual life time must be independent of t, i.e. it coincides with the distribution of B_0 and hence of X_0. Regarding the limit given in (6.9), we first guess that it satisfies

$$\mathbb{P}(X_0 \leq x) = \frac{1}{m} \int_0^x (1 - F(y))\, dy \qquad (6.11)$$

for all $x \geq 0$, where F denotes the distribution function of X_1 and further $m = \mathbb{E}(X_1) < \infty$. Indeed we can show

Theorem 6.23 *For a renewal process $\mathcal{N}$ defined by (6.11) the following properties hold:*
(1) $\mathbb{E}(N_t) = t/m$ for all $t \geq 0$
(2) $\mathbb{P}(B_t \leq x) = m^{-1} \int_0^x (1 - F(y))\, dy$ for all $t \geq 0$
(3) $\mathcal{N}$ has stationary increments.

Proof: (1) The distribution G of X_0 has a density $g(t) = \frac{1}{m}(1 - F(t))$ Hence the Laplace–Stieltjes transform (LST) of G is

$$\tilde{G}(s) = \int_0^\infty e^{-st} \frac{1}{m}(1 - F(t))\, dt = \frac{1}{m}\left(\int_0^\infty e^{-st}\, dt - \int_0^\infty e^{-st} F(t)\, dt \right)$$
$$= \frac{1}{m}\left(\frac{1}{s} - \frac{1}{s}\int_0^\infty e^{-st}\, dF(t) \right) = \frac{1 - \tilde{F}(s)}{sm}$$

with $\tilde{F}(s)$ denoting the LST of F. According to (6.1) we have the representation $\mathbb{E}(N_t) = G * \sum_{n=0}^\infty F^{*n}(t)$ for all $t \geq 0$. Hence the LST of $M(t) := \mathbb{E}(N_t)$ is given by

$$\tilde{M}(s) = \frac{\tilde{G}(s)}{1 - \tilde{F}(s)} = \frac{1}{sm}$$

for all $s > 0$, and thus coincides with the LST of the measure dx/m. Since the LST uniquely determines a function on $[0, \infty[$, this proves the first statement.

(2) The joint distributions

$$\mathbb{P}(B_t > x, N_t = 0) = 1 - G(t + x)$$

$$\mathbb{P}(B_t > x, N_t = n) = \int_0^\infty \mathbb{P}(B_t > x, N_t = n | S_n = y) \, dG * F^{*n-1}(y)$$

$$= \int_0^t (1 - F(t + x - y)) \, dG * F^{*n-1}(y)$$

for $n \geq 1$ are immediate from the definition. Abbreviating $F^c(x) := 1 - F(x)$, $G^c(x) := 1 - G(x)$, and denoting $M(t) := \mathbb{E}(N_t)$, we can write

$$\mathbb{P}(B_t > x) = \sum_{n=0}^\infty \mathbb{P}(B_t > x, N_t = n)$$

$$= G^c(t + x) + \sum_{n=1}^\infty \int_0^t F^c(t + x - y) \, dG * F^{*n-1}(y)$$

$$= G^c(t + x) + \int_0^t F^c(t + x - y) \, d \left(\sum_{n=1}^\infty G * F^{*n-1} \right) (y)$$

$$= G^c(t + x) + \int_0^t F^c(t + x - y) \, dM(y)$$

Using statement (1) and the definition of G, we obtain

$$\mathbb{P}(B_t > x) = 1 - \frac{1}{m} \int_0^{t+x} (1 - F(y)) \, dy + \frac{1}{m} \int_0^t (1 - F(t + x - y)) \, dy$$

$$= 1 - \frac{1}{m} \int_0^x (1 - F(y)) \, dy$$

which proves the second statement.

(3) The difference $N_{t+s} - N_s$ simply counts the number N_t' of events in time t of the renewal process $\mathcal{N}'$ with the same distribution F of X_1 but a delay $X_0' \sim B_s$. Now statement (2) shows that $X_0 \sim B_s = B_0$. Hence we obtain $N_t' = N_t = N_{t+s} - N_s$ in distribution, which was to be proven.
□

Because of the results above a renewal process which satisfies condition (6.11) is called **stationary renewal process**. As one would expect, also the mean residual life time $\mathbb{E}(B_t)$ of a stationary renewal process coincides with the limit of the mean residual life time of an ordinary renewal process:

Lemma 6.24 *For a non–negative random variable X the nth moment can be expressed by*

$$\mathbb{E}(X^n) = \int_0^\infty \mathbb{P}(X > x) \cdot n x^{n-1} \, dx$$

Proof: This follows simply by writing

$$\mathbb{E}(X^n) = \int_0^\infty \mathbb{P}(X^n > z)\, dz = \int_0^\infty \mathbb{P}(X > \sqrt[n]{z})\, dz$$

and substituting $x = \sqrt[n]{z}$ with $nx^{n-1}\, dx = dz$.
$\square$

Theorem 6.25 *For a stationary renewal process with $\mathbb{E}(X_1^2) < \infty$ the mean residual life time is given by*

$$\mathbb{E}(B_t) = \frac{\mathbb{E}(X_1^2)}{2m}$$

independently of $t \geq 0$.

Proof: Using part (2) of theorem 6.23, we obtain

$$\mathbb{E}(B_t) = \int_0^\infty \mathbb{P}(B_t > y)\, dy = \frac{1}{m} \int_{y=0}^\infty \int_{x=y}^\infty (1 - F(x))\, dx\, dy$$

$$= \frac{1}{m} \int_{x=0}^\infty \int_{y=0}^x (1 - F(x))\, dy\, dx = \frac{1}{m} \int_{x=0}^\infty \mathbb{P}(X_1 > x) \cdot x\, dx$$

and the statement follows from lemma 6.24.
$\square$

Example 6.26 Waiting time at a bus stop
Consider a bus stop where buses are scheduled to arrive in intervals of length T. However, due to traffic variations the real inter–arrival times are uniformly distributed within intervals $[T - a, T + a]$ with some $a > 0$. Now suppose that somebody arrives at the bus stop "at random", i.e. without knowing the bus schedule. Then we can model the mean waiting time for the next bus by the mean residual life time $\mathbb{E}(B_t)$ in a stationary renewal process with distribution $X_1 \sim U(T - a, T + a)$. We obtain

$$\mathbb{E}(X_1^2) = \frac{1}{2a} \int_{T-a}^{T+a} x^2\, dx = \frac{1}{2a} \cdot \frac{1}{3} \left(6T^2 a + 2a^3\right) = T^2 + \frac{a^2}{3}$$

and by theorem 6.25

$$\mathbb{E}(B_t) = \frac{T^2 + \frac{a^2}{3}}{2 \cdot T} = \frac{T}{2} + \frac{a^2}{6 \cdot T}$$

Thus the mean waiting time for random inter–arrival times (meaning $a > 0$) is longer than it would be for deterministic ones (namely $T/2$). This phenomenon is called the **waiting time paradox**.

5.   Renewal Reward Processes

Consider an ordinary renewal process where for every renewal interval X_n there is a real–valued random variable Y_n, called the nth **reward**, which may depend on X_n. If the pairs $(X_n, Y_n), n \in \mathbb{N}_0$ are iid, then the two–dimensional stochastic chain $((X_n, Y_n) : n \in \mathbb{N}_0)$ is called an ordinary **renewal reward process**. The random variable

$$Y(t) = \sum_{n=0}^{N_t-1} Y_n$$

is called the **total reward** until time t.

Theorem 6.27 *If $\mathbb{E}(|Y_1|)$ and $m = \mathbb{E}(X_1)$ are finite, then*

$$\lim_{t \to \infty} \frac{Y(t)}{t} = \frac{\mathbb{E}(Y_1)}{m}$$

holds with probability one. If there is further a constant $c \in \mathbb{R}$ with $Y_1 > c$ almost certainly, then

$$\lim_{t \to \infty} \frac{\mathbb{E}(Y(t))}{t} = \frac{\mathbb{E}(Y_1)}{m}$$

Proof: The first statement follows from

$$\frac{Y(t)}{t} = \frac{\sum_{n=0}^{N_t-1} Y_n}{N_t} \cdot \frac{N_t}{t}$$

as the first factor tends to $\mathbb{E}(Y_1)$ by the strong law of large numbers and the second tends to $1/m$ according to theorem 6.4.

For the second statement, we can assume without loss of generality that Y_1 is positive almost certainly, since otherwise we consider $Z_n := Y_n + c$ instead. N_t is a stopping time for the sequence $(Y_n : n \in \mathbb{N}_0)$, as $\{N_t \le n\}$ is independent of $(X_{n+k} : k \in \mathbb{N})$ and thus independent of $(Y_{n+k} : k \in \mathbb{N})$. Hence we can apply Wald's lemma, which yields

$$\mathbb{E}\left(\sum_{n=0}^{N_t-1} Y_n\right) = \mathbb{E}\left(\sum_{n=0}^{N_t} Y_n\right) - \mathbb{E}(Y_{N_t}) = R(t) \cdot \mathbb{E}(Y_1) - \mathbb{E}(Y_{N_t})$$

and thus

$$\frac{\mathbb{E}(Y(t))}{t} = \frac{R(t)}{t} \cdot \mathbb{E}(Y_1) - \frac{\mathbb{E}(Y_{N_t})}{t}$$

for all $t > 0$. Because of $\lim_{t\to\infty} R(t)/t = 1/m$ it now remains to show that $\lim_{t\to\infty} \mathbb{E}(Y_{N_t})/t = 0$. To this aim we condition on X_0 to obtain

$$g(t) := \mathbb{E}(Y_{N_t}) = \int_0^\infty \mathbb{E}(Y_{N_t}|X_0 = u) \ dF(u)$$

$$= \int_0^t \mathbb{E}(Y_{N_t}|X_0 = u) \ dF(u) + \int_t^\infty \mathbb{E}(Y_{N_t}|X_0 = u) \ dF(u)$$

for all $t > 0$. Abbreviating the latter integral by $h(t)$ and recognizing that $\mathbb{E}(Y_{N_t}|X_0 = u) = g(t - u)$, we obtain the renewal equation

$$g(t) = \int_0^t g(t - u) \ dF(u) + h(t)$$

Theorem 6.9 yields the unique solution

$$g(t) = h(t) + \int_0^t h(t - u) \ dR(u)$$

for all $t > 0$. As $X_0 > t$ implies $N_t = 0$, we know further that

$$h(t) = \int_t^\infty \mathbb{E}(Y_0|X_0 = u) \ dF(u) \le \mathbb{E}(Y_0) < \infty$$

and $h(t) \to 0$ as $t \to \infty$. This means that for any given $\varepsilon > 0$ there is a $T > 0$ such that $|h(t)| < \varepsilon$ for all $t \ge T$. Using this we obtain

$$\frac{|g(t)|}{t} \le \frac{|h(t)|}{t} + \frac{1}{t} \int_0^{t-T} |h(t - u)| \ dR(u) + \frac{1}{t} \int_{t-T}^t |h(t - u)| \ dR(u)$$

$$\le \frac{\varepsilon}{t} + \varepsilon \cdot \frac{R(t - T)}{t} + \mathbb{E}(Y_0) \cdot \frac{R(t) - R(t - T)}{t}$$

for all $t > T$. For $t \to \infty$ the right–hand side tends to ε/m by the elementary renewal theorem, as $R(t) - R(t - T)$ is bounded by $R(T) + 1$ (see the proof of the key renewal theorem). This completes the proof, as ε can be chosen arbitrarily small.
□

Example 6.28 Machine maintenance
Consider a machine that is prone to failure and may be either repaired or re-placed by a new machine. Let X_n denote the run time of the machine after the $n - 1$st failure and assume $\lambda := \mathbb{E}(X_1) < \infty$. Since the state of the machine after the nth repair is usually worse than after the $n - 1$st repair, we model this by the assumption that $(a^{n-1}X_n : n \in \mathbb{N})$ with $a \ge 1$ forms a renewal process. In particular, the $X_n, n \in \mathbb{N}$ are independent random variables. The sequence

$(X_n : n \in \mathbb{N})$ is called a non–increasing **geometric process** with parameter a. The reward rate for the machine running is $r = 1$.

The duration of the nth repair is denoted by Y_n, $n \in \mathbb{N}$, with the assumption $\mu := \mathbb{E}(Y_1) < \infty$. As the machine becomes more and more difficult to repair, we assume that $(b^{n-1}Y_n : n \in \mathbb{N})$ with $b \leq 1$ forms a renewal process. The sequence $(Y_n : n \in \mathbb{N})$ is called a non–decreasing geometric process with parameter b. Again this implies that the $Y_n, n \in \mathbb{N}$ are independent random variables. Furthermore we assume that $\{X_n, Y_n : n \in \mathbb{N}\}$ is an independent set of random variables. The cost (i.e. the negative reward) rate for the repair of the machine is denoted by $c_1 > 0$.

Instead of repairing the machine after a failure, we can choose to replace it by a new machine. This incurs a fixed cost $c_2 > c_1$. Given this information, we want to determine the long–run expected reward per unit time for the machine. This depends on the variable $N \in \mathbb{N}$ which indicates the policy that a machine is replaced after the Nth failure.

Clearly the replacement times $(T_n : n \in \mathbb{N}_0)$ with $T_0 := 0$ form an ordinary renewal process and the reward of a machine (i.e. between replacement times) is independent from the rewards and life times of other machines. Denote the life time and the reward of the nth machine by $L_n := T_n - T_{n-1}$ and R_n, respectively. Then $((L_n, R_n) : n \in \mathbb{N})$ is a renewal reward process and the long–run expected reward per unit time is given by

$$R(N) = \frac{\mathbb{E}(R_1)}{\mathbb{E}(L_1)} = \frac{\lambda \sum_{k=1}^{N} a^{-(k-1)} - c_1 \cdot \mu \sum_{k=1}^{N-1} b^{-(k-1)} - c_2}{\lambda \sum_{k=1}^{N} a^{-(k-1)} + \mu \sum_{k=1}^{N-1} b^{-(k-1)}}$$

according to theorem 6.27. In order to find the optimal replacement policy, this equation can now be used to determine the value N which maximizes the expected reward rate $R(N)$.

Notes

A classical presentation of renewal theory is chapter 11 in Feller [35]. The presentation in this chapter is largely adapted to Ross [74, 75] as well as Karlin and Taylor [46]. The concept of regenerative processes has been developed by Feller and Smith [80, 81]. Example 6.28 is taken from Lam Yeh [51].

Exercise 6.1 Prove theorem 6.3.

Exercise 6.2 In the proof for Wald's lemma 6.11 we have used the relation $\mathbb{E}(S) = \sum_{n=0}^{\infty} \mathbb{P}(S > n)$, and in theorem 6.18 $\mathbb{E}(F) = \int_0^{\infty}(1 - F(y)) \, dy$. Give a proof for these equations.

Exercise 6.3 Show for a Poisson process $\mathcal{N}$ with intensity $\lambda > 0$ that

$$\mathbb{P}(N_t = k) = \frac{(\lambda t)^k}{k!} e^{-\lambda t}$$

for all $t \geq 0$ and $k \in \mathbb{N}_0$, and $\mathbb{E}(N_t) = \lambda \cdot t$.

Exercise 6.4 A plumber receives orders at time intervals which are distributed exponentially with parameter λ. As soon as he has received an order he goes to work, which takes an exponentially distributed time with parameter μ. During work he cannot receive any orders. Assume that at time zero the plumber is working. Give a model of the plumber receiving orders in terms of a renewal process and determine the density of the renewal intervals' distribution.

Exercise 6.5 An intelligence agency eavesdrops on telephone calls automatically. If there occurs a suspicious sequence of words, a closer investigation is initiated. The probabilitiy for such a sequence is one in a thousand for every call. The length of a call is distributed exponentially with a mean of 20 seconds. How long is the expected amount of time before a closer investigation begins? Use Wald's lemma.

Exercise 6.6 Let $\mathcal{N} = (N_t : t \geq 0)$ denote an ordinary renewal process with $X_1 \sim F$. Show that the current life time X_{N_t} satisfies

$$\mathbb{P}(X_{N_t} > x) \geq 1 - F(x)$$

for all $x \geq 0$.

Exercise 6.7 Give an example which shows why we need to assume in Blackwell's theorem that the distribution of the renewal intervals is not lattice.

Exercise 6.8 Prove lemma 6.16.

Exercise 6.9 Show that the age A_t of a stationary renewal process is distributed as

$$\mathbb{P}(A_t \leq x) = \frac{1}{m} \int_0^x (1 - F(y)) \, dy$$

independently of $t \geq 0$.

Exercise 6.10 Show that for an ordinary renewal process with $\mathbb{E}(X_1^2) < \infty$ and $m := \mathbb{E}(X_1)$ the limit

$$\lim_{t \to \infty} \mathbb{E}(A_t) = \frac{\mathbb{E}(X_1^2)}{2m}$$

holds.

Exercise 6.11 Passengers arrive at a train station according to an ordinary renewal process with rate $1/m$. As soon as there are N passengers waiting, the train departs. The cost for the ticket per passenger is C. Assume that the railway company reimburses every passenger for the waiting time by an amount of W per time unit that the passenger had to wait. Determine the minimal value for C such that this train connection will be profitable in the long run.

Exercise 6.12 A **delayed renewal reward process** is defined as a stochastic chain $((X_n, Y_n) : n \in \mathbb{N}_0)$ for which $((X_n, Y_n) : n \in \mathbb{N})$ is an ordinary renewal reward process and $X_0 \geq 0$. The pair (X_0, Y_0) may have a different distribution than (X_1, Y_1). Prove the statement of theorem 6.27 for a delayed renewal reward process that satisfies $\mathbb{E}(X_0) < \infty$ and $\mathbb{E}(|Y_0|) < \infty$.

Chapter 7

MARKOV RENEWAL THEORY

1.   Regenerative Processes

Let $\mathcal{Y} = (Y_t : t \geq 0)$ denote a stochastic process on a discrete state space E with right–continuous paths. Further let T denote a random variable with values in $[0, \infty]$ such that the condition

$$\mathbb{P}(T \leq t | \mathcal{Y}) = \mathbb{P}(T \leq t | Y_s : s \leq t) \tag{7.1}$$

holds for all $t \in \mathbb{R}_0^+$. Such a random variable is called a (continuous) **stopping time** for the process $\mathcal{Y}$. As in the analogue for discrete time, the defining condition means that the probability for the event $\{T \leq t\}$ depends only on the evolution of the process until Y_t. In other words, the determination of a stopping time does not require any knowledge of the future.

If there is a sequence $\mathcal{T} = (T_n : n \in \mathbb{N}_0)$ of stopping times for $\mathcal{Y}$ with $T_0 := 0$ and $T_{n+1} > T_n$ for all $n \in \mathbb{N}_0$ such that $\mathcal{T}$ defines an ordinary renewal process, and if further

$$\mathbb{P}(Y_{T_n+t_1} = j_1, \dots, Y_{T_n+t_k} = j_k | Y_u : u \leq T_n) = \mathbb{P}(Y_{t_1} = j_1, \dots, Y_{t_k} = j_k)$$

for all $k \in \mathbb{N}$, $t_1, \dots, t_k \geq 0$ and $n \in \mathbb{N}_0$ holds, then $\mathcal{Y}$ is called a **regenerative process**. The T_n are called **regeneration time**s and the defining condition above is called **regeneration property**. The interval $[T_{n-1}, T_n[$ is called the nth **regeneration cycle**.

Example 7.1 M/G/k Queue
The **M/G/k queue** has a Poisson arrival process and k servers with general service time distribution. Whenever the queue is empty, all servers are idle and

only the arrival process has an effect on the future. Thus the system process regenerates at the points T_n of the system becoming idle for the nth time. The durations $T_{n+1} - T_n$ between these points are iid. Hence the M/G/k system process is a regenerative process.

Theorem 7.2 *If T_1 is not lattice and $\mathbb{E}(T_1) = m < \infty$ holds and if the function $K_j(t) := \mathbb{P}(T_1 > t, Y_t = j)$ is Riemann integrable, then*

$$\pi_j := \lim_{t \to \infty} \mathbb{P}(Y_t = j) = \frac{1}{m} \int_0^\infty K_j(t) \, dt$$

for all $j \in E$.

Proof: Let F denote the distribution function of T_1. By conditioning on the first regeneration time T_1, we obtain the equation

$$\mathbb{P}(Y_t = j) = \mathbb{P}(T_1 > t, Y_t = j) + \int_0^t \mathbb{P}(Y_t = j | T_1 = s) \, dF(s)$$

$$= \mathbb{P}(T_1 > t, Y_t = j) + \int_0^t \mathbb{P}(Y_{t-s} = j) \, dF(s)$$

where the second equality is due to the regeneration property. The function $K_j(t) = \mathbb{P}(T_1 > t, Y_t = j)$ is non–negative and bounded by $\mathbb{P}(T_1 > t)$, which in turn is Lebesgue integrable and non–increasing. By assumption $K_j(t)$ is Riemann integrable. Hence lemma 6.16 yields that $K_j(t)$ is directly Riemann integrable. Thus the key renewal theorem 6.17 applies and yields the statement. $\square$

Introduce a real–valued function $f : E \to \mathbb{R}$ on the state space of the process $\mathcal{Y}$. The value $f(i)$ can be interpreted as a reward rate which is incurred in state $i \in E$.

Theorem 7.3 *If $\mathbb{E}(T_1) < \infty$ and f is bounded, then*

$$\lim_{t \to \infty} \frac{1}{t} \int_0^t f(Y_s) \, ds = \frac{\mathbb{E} \int_0^{T_1} f(Y_t) \, dt}{\mathbb{E}(T_1)}$$

holds with probability one. If further $\mathbb{E}(T_1^2) < \infty$, then

$$\lim_{t \to \infty} \mathbb{E} \left(\frac{1}{t} \int_0^t f(Y_s) \, ds \right) = \frac{\mathbb{E} \int_0^{T_1} f(Y_t) \, dt}{\mathbb{E}(T_1)}$$

Proof: Define $X_n := T_{n+1} - T_n$ and $Z_n := \int_{T_n}^{T_{n+1}} f(Y_s) \, ds$ for all $n \in \mathbb{N}_0$. Since $\mathcal{Y}$ is regenerative, the chain $((X_n, Z_n) : n \in \mathbb{N}_0)$ is a renewal reward process, with $N_t := \max\{n \in \mathbb{N}_0 : T_n \leq t\}$ and $Z(t) := \sum_{n=0}^{N_t-1} Z_n$ defined as usual. We can write

$$\int_0^t f(Y_s)\,ds = Z(t) + \int_{T_{N_t}}^t f(Y_s)\,ds \tag{7.2}$$

for all $t \geq 0$. Since $Z(t)/t$ converges to the fraction on the right–hand side of the statement (see theorem 6.27), it remains to show that $t^{-1} \int_{T_{N_t}}^t f(Y_s)\,ds$ tends to zero as $t \to \infty$. We obtain

$$\frac{1}{t} \int_{T_{N_t}}^t f(Y_s)\,ds = \frac{1}{N_t} \int_{T_{N_t}}^t f(Y_s)\,ds \cdot \frac{N_t}{t} \leq \frac{X_{N_t}}{N_t} \cdot \sup_{i \in E} |f(i)| \cdot \frac{N_t}{t}$$

$$\to \lim_{n \to \infty} \frac{X_n}{n} \cdot \sup_{i \in E} |f(i)| \cdot \lim_{t \to \infty} \frac{N_t}{t}$$

as $t \to \infty$. According to the strong law of large numbers, we know that $\sum_{k=1}^n X_k/n \to m < \infty$ almost certainly and hence $X_n/n \to 0$ as $n \to \infty$. This and theorem 6.4 complete the proof for the first statement.

The same partition (7.2) and theorem 6.27 show that for the second statement it remains to show that $t^{-1}\mathbb{E}\left(\int_{T_{N_t}}^t f(Y_s)\,ds\right) \to 0$ as $t \to \infty$. However, this follows from

$$\int_{T_{N_t}}^t f(Y_s)\,ds \leq A_t \cdot \sup_{i \in E} |f(i)|$$

and $\lim_{t \to \infty} \mathbb{E}(A_t) = \mathbb{E}(X_1^2)/(2m) < \infty$ by exercise 6.10 and the assumption that $\mathbb{E}(T_1^2)$ be finite.

$\square$

Theorem 7.4 *If T_1 is not lattice and $\mathbb{E}(T_1)$ as well as $\mathbb{E}\left|\int_0^{T_1} f(Y_t)\,dt\right|$ are finite, then*

$$\frac{\mathbb{E}\int_0^{T_1} f(Y_t)\,dt}{\mathbb{E}(T_1)} = \sum_{j \in E} \pi_j \cdot f(j)$$

with π_j as defined in theorem 7.2. If T_1 is not lattice and $\mathbb{E}(T_1) < \infty$, then

$$\pi_j = \lim_{t \to \infty} \frac{1}{t} \int_0^t \mathbf{1}_{\{Y_s = j\}}\,ds$$

holds with probability one for all $j \in E$. This means that the limiting probability π_j of j equals the asymptotic proportion of time spent in state j for every path.

Proof: The Lebesgue construction of an integral yields

$$\int_0^{T_1} f(Y_t)\, dt = \sum_{j \in E} f(j) \cdot \int_0^{T_1} \mathbf{1}_{\{Y_t = j\}}\, dt$$

and after taking expectations we obtain

$$\mathbb{E} \int_0^{T_1} f(Y_t)\, dt = \sum_{j \in E} f(j) \cdot \int_0^{\infty} K_j(t)\, dt$$

with $K_j(t) = \mathbb{P}(T_1 > t, Y_t = j)$. Now the first statement follows from theorem 7.2. The second statement follows from the first one and theorem 7.3 for $f(Y_t) := \mathbf{1}_{\{Y_t = j\}}$.

□

2.   Semi–Markov Processes

In this section we will introduce a special class of regenerative processes which is very useful for the analysis of many queueing systems.

Let E denote a countable state space. For every $n \in \mathbb{N}_0$, let X_n denote a random variable on E and T_n a random variable on $\mathbb{R}_0^+$ such that $T_0 := 0$, $T_n < T_{n+1}$ for all $n \in \mathbb{N}_0$, and $\sup_{n \to \infty} T_n = \infty$ almost surely. Define the process $\mathcal{Y} = (Y_t : t \in \mathbb{R}_0^+)$ by

$$Y_t := X_n \qquad \text{for} \quad T_n \leq t < T_{n+1}$$

for all $t \geq 0$. If

$$\mathbb{P}(X_{n+1} = j, T_{n+1} - T_n \leq u | X_0, \ldots, X_n, T_0, \ldots, T_n)$$
$$= \mathbb{P}(X_{n+1} = j, T_{n+1} - T_n \leq u | X_n) \quad (7.3)$$

holds for all $n \in \mathbb{N}_0$, $j \in E$, and $u \in \mathbb{R}_0^+$, then $\mathcal{Y}$ is called a **semi–Markov process** on E. The sequence $(\mathcal{X}, \mathcal{T}) = ((X_n, T_n) : n \in \mathbb{N}_0)$ of random variables is called the **embedded Markov renewal chain**. We will treat only **homogeneous** semi–Markov processes, i.e. those for which

$$F_{ij}(t) := \mathbb{P}(X_{n+1} = j, T_{n+1} - T_n \leq t | X_n = i)$$

is independent of n. For all $i, j \in E$, the functions $t \to F_{ij}(t)$ are assumed non–lattice.

By definition a semi–Markov process is a pure jump process. Thus the sample paths are step functions:

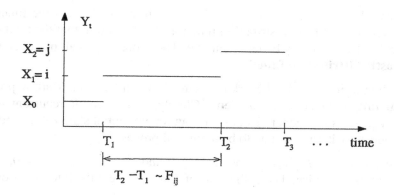

Figure 7.1. Typical path of a semi–Markov process

By construction, the semi–Markov process $\mathcal{Y}$ is determined by the embedded Markov renewal chain $(\mathcal{X}, \mathcal{T})$ and vice versa.

Remark 7.5 Let $\mathcal{Y}$ denote an homogeneous Markov process with discrete state space E and parameters λ_i, $i \in E$, for the exponential holding times. The embedded Markov chain $\mathcal{X}$ of $\mathcal{Y}$ shall have transition matrix $P = (p_{ij})_{i,j \in E}$. Then $\mathcal{Y}$ is a semi–Markov process with

$$F_{ij}(t) = p_{ij} \cdot \left(1 - e^{-\lambda_i \cdot t}\right)$$

for all $i, j \in E$. Thus for a Markov process the distribution of $T_{n+1} - T_n$ is exponential and independent of the state entered at time T_{n+1}. These are the two features for which the semi–Markov process is a generalization of the Markov process on a discrete state space.

Theorem 7.6 *Let $\mathcal{Y}$ be a semi–Markov process with embedded Markov renewal chain $(\mathcal{X}, \mathcal{T})$. Then $\mathcal{X} = (X_n : n \in \mathbb{N}_0)$ is a Markov chain.*

Proof: From the condition (7.3) we obtain for every $n \in \mathbb{N}_0$

$$\begin{aligned}
\mathbb{P}(X_{n+1} = j | X_0, \dots, X_n) &= \mathbb{P}(X_{n+1} = j, T_{n+1} - T_n < \infty | X_0, \dots, X_n) \\
&= \mathbb{P}(X_{n+1} = j, T_{n+1} - T_n < \infty | X_n) \\
&= \mathbb{P}(X_{n+1} = j | X_n)
\end{aligned}$$

since all T_n are finite by definition.
$\square$

We denote the transition matrix of $\mathcal{X}$ by $P = (p_{ij})_{i,j \in E}$. Then the relation

$$p_{ij} := \mathbb{P}(X_{n+1} = j | X_n = i) = \lim_{t \to \infty} F_{ij}(t)$$

obviously holds for all $i, j \in E$. This means in particular that the functions $F_{ij}(t)$ are distinct from distribution functions in the feature that the total mass distributed by them may be less than one. Therefore they shall be called **sub–stochastic distribution functions**.

According to its embedded Markov chain $\mathcal{X}$, we call a semi–Markov process **irreducible, recurrent** or **transient**. Clearly, an irreducible recurrent semi–Markov process is regenerative, as one can fix any initial state $i \in E$ and find the times of visiting this state to be a renewal process.

Define $G_{ij}(t) := F_{ij}(t)/p_{ij}$ for all $t \geq 0$ and $i, j \in E$ if $p_{ij} > 0$, while $G_{ij}(t) := 0$ otherwise. The definitions of P and F yield the interpretation

$$G_{ij}(t) = \mathbb{P}(T_{n+1} - T_n \leq t | X_n = i, X_{n+1} = j)$$

which in turn yields

Theorem 7.7 *Let $\mathcal{Y}$ denote a semi–Markov process with state space E and embedded Markov renewal chain $(\mathcal{X}, \mathcal{T})$. Then*

$$\mathbb{P}(T_1 - T_0 \leq u_1, \ldots, T_n - T_{n-1} \leq u_n | X_0, \ldots, X_n)$$
$$= G_{X_0, X_1}(u_1) \ldots G_{X_{n-1}, X_n}(u_n)$$

for all $n \in \mathbb{N}$, meaning that the increments $T_1 - T_0, \ldots, T_n - T_{n-1}$ are conditionally independent, given the values $X_0, \ldots, X_n$.

Remark 7.8 If the state space E is trivial, i.e. consisting of only one element, then these increments are even iid. In this case, $\mathcal{T} = (T_n : n \in \mathbb{N}_0)$ is a renewal process. This property and theorem 7.6 justify the name Markov renewal theory for the study of semi–Markov processes, as the latter generalize Markov processes and renewal processes at the same time.

2.1   Transient distributions

For Markov chains and Markov processes we have been able to give formulae for the transition matrices. Because of the Markov property, this in turn determined all finite–dimensional marginal distributions and thus the complete distribution of the process. In the case of a semi–Markov process, we cannot give as much information. However, what we can derive are the transition probabilities starting from a regeneration time. Since $T_0 := 0$ is a deterministic regeneration point, this yields together with a given initial distribution the one–dimensional marginal distributions at any given time. These shall be called **transient distributions**.

In order to determine the transient distributions of a semi–Markov process, we will view the collection $F = (F_{ij})_{i,j \in E}$ of sub–stochastic distribution functions as a matrix with entries being functions instead of numbers. This matrix contains all stochastic laws for the construction of the semi–Markov process $\mathcal{Y}$. Therefore it shall be called the **characterizing matrix** of $\mathcal{Y}$.

We define a **matrix convolution** of two such matrices F and G by the entries

$$(F * G)_{ij}(t) := \sum_{k \in E} \int_0^t G_{kj}(t - u) \, dF_{ik}(u)$$

for all $i, j \in E$ and $t \geq 0$. Based on this definition, we define the matrix convolutional powers by $F^{*0} := I_E$, denoting the identity matrix on E, and by recursion $F^{*n+1} = F^{*n} * F$. Now we can state the following formula for the transient distributions of a semi–Markov process:

Theorem 7.9 *Let $\mathcal{Y}$ denote a semi–Markov process with characterizing matrix F, and π any initial distribution of $\mathcal{Y}$. Then the transient distribution of $\mathcal{Y}^\pi$ at any time t is given by $\mathbb{P}(Y_t^\pi = j) = \sum_{i \in E} \pi_i P_{ij}(t)$ with*

$$P_{ij}(t) = \sum_{n=0}^{\infty} \int_0^t \left(1 - \sum_{k \in E} F_{jk}(t - u)\right) dF_{ij}^{*n}(u)$$

Proof: This expression is obtained by conditioning on the number n of renewal intervals that have passed until time t.
□

2.2   Asymptotic behaviour

Next we want to examine the asymptotic behaviour of the transient distributions, i.e. we want to determine the limits $\lim_{t \to \infty} \mathbb{P}(Y_t = j)$ for all $j \in E$. This will be achieved by specifying the results which have already been obtained for regenerative processes.

If we want to use theorem 7.2, then we need the information for the respective regenerative process with embedded renewals being the visits to any fixed state $j \in E$. Define

$$m_{ij} := \mathbb{E}(T_1 \cdot \mathbf{1}_{X_1=j} | X_0 = i) = \int_0^{\infty} t \, dF_{ij}(t)$$

$$m_i := \mathbb{E}(T_1 | X_0 = i) = \sum_{j \in E} m_{ij} = \int_0^{\infty} \left(1 - \sum_{k \in E} F_{ik}(t)\right) dt$$

for all $i, j \in E$. Further define

$$\tau_j := \min\{T_n : X_n = j, n \in \mathbb{N}\} \quad \text{and} \quad \mu_{ij} := \mathbb{E}(\tau_j | X_0 = i) \qquad (7.4)$$

for all $i, j \in E$. The random variable τ_j is called **first return time** to state j.

Now consider any Markov renewal time T_n. If the Markov chain $\mathcal{X}$ is irreducible and positive recurrent with stationary distribution $\nu = (\nu_i : i \in E)$, then we would expect a proportion ν_j of sample paths with transition into j at time T_n. Furthermore, m_j represents the mean time spent in state j until the next transition happens at time T_{n+1}. Therefore, if there is an asymptotic distribution $\pi_j = \lim_{t\to\infty} \mathbb{P}(Y_t = j)$, we would expect it to be proportional to $\nu_j \cdot m_j$, i.e.

$$\pi_j = \frac{\nu_j \cdot m_j}{\sum_{i\in E} \nu_i \cdot m_i}$$

We will prove this by examining the above mentioned embedded regenerative process of visits to state j.

Lemma 7.10 *The relation*

$$\mu_{ij} = m_i + \sum_{k\neq j} p_{ik}\mu_{kj}$$

holds for all $i, j \in E$.

Proof: Conditioning on the state $X_1 = k$ at time T_1, we can write

$$\mu_{ij} = \sum_{k\in E} \mathbb{E}(\tau_j \cdot \mathbf{1}_{X_1=k}|X_0 = i) = \sum_{k\neq j}(p_{ik}\mu_{kj} + m_{ik}) + m_{ij}$$

$$= \sum_{k\in E} m_{ik} + \sum_{k\neq j} p_{ik} \cdot \mu_{kj} = m_i + \sum_{k\neq j} p_{ik}\mu_{kj}$$

which is the statement.

$\square$

Lemma 7.11 *Let $\mathcal{Y}$ denote a semi–Markov process with embedded Markov renewal chain $(\mathcal{X}, \mathcal{T})$. Assume that $\mathcal{X}$ is positive recurrent and denote its stationary distribution by $\nu = \nu P$. Further assume that $\sum_{i\in E} \nu_i m_i < \infty$. Then the **mean recurrence time** of a state $j \in E$ can be expressed by*

$$\mu_{jj} = \mathbb{E}(\tau_j|X_0 = j) = \frac{1}{\nu_j} \sum_{i\in E} \nu_i m_i$$

for all $j \in E$.

Proof: We multiply both sides of lemma 7.10 by ν_i and sum up over all $i \in E$. Then we obtain

$$\sum_{i \in E} \nu_i \mu_{ij} = \sum_{i \in E} \nu_i m_i + \sum_{i \in E} \nu_i \sum_{k \neq j} p_{ik} \mu_{kj} = \sum_{i \in E} \nu_i m_i + \sum_{k \neq j} \mu_{kj} \sum_{i \in E} \nu_i p_{ik}$$

$$= \sum_{i \in E} \nu_i m_i + \sum_{k \neq j} \nu_k \mu_{kj}$$

which implies

$$\nu_j \mu_{jj} = \sum_{i \in E} \nu_i m_i$$

and thus the statement.
□

Theorem 7.12 *Let $\mathcal{Y}$ denote a semi–Markov process with embedded Markov renewal chain $(\mathcal{X}, \mathcal{T})$ and characterizing matrix F. Assume that $\mathcal{X}$ is irreducible and positive recurrent and $\nu = \nu P$ is the stationary distribution of its transition matrix P. Further assume that $\sum_{i \in E} \nu_i m_i < \infty$. Then the limits*

$$\pi_j := \lim_{t \to \infty} \mathbb{P}(Y_t = j) = \frac{\nu_j m_j}{\sum_{i \in E} \nu_i m_i}$$

hold for all $j \in E$, independent of the initial distribution.

Proof: Since the times of successive visits to state j form a (possibly delayed) renewal process, the process $\mathcal{Y}$ is regenerative. Since all functions F_{ij} are assumed non–lattice, the regeneration cycles of $\mathcal{Y}$ are not lattice either. Thus we can apply theorem 7.2 (in the form of exercise 7.1), which yields

$$\lim_{t \to \infty} \mathbb{P}(Y_t = j) = \frac{m_j}{\mu_{jj}}$$

Now lemma 7.11 leads to the statement.
□

Example 7.13 This limit theorem finally suffices for an application to Markov processes. The two limit theorems (3.6) and (3.7) follow from the interpretation of a Markov process as a special semi–Markov process. For a Markov process, the mean holding times in a state i are given by

$$m_i = \int_0^\infty \mathbb{P}(T_1 > t | Y_0 = i)\, dt = \int_0^\infty e^{-\lambda_i \cdot t}\, dt = \frac{1}{\lambda_i}$$

for all $i \in E$. Since $\lambda_i \geq \check{\lambda} > 0$ for all $i \in E$, we know that $\sum_{i \in E} \nu_i m_i < \infty$. Hence we obtain

$$\lim_{t \to \infty} P_{ij}(t) = \lim_{t \to \infty} \mathbb{P}(Y_t = j) = \frac{\nu_j / \lambda_j}{\sum_{i \in E} \nu_i / \lambda_i}$$

as given in equation (3.5).

3.   Semi–regenerative Processes

Semi–Markov processes play a similar role for the analysis of a more general class of processes that renewal processes have played for the analysis of regenerative processes. These more general processes are called semi–regenerative and shall be defined as follows:

Let $\mathcal{Z} = (Z_t : t \in \mathbb{R}_0^+)$ denote a stochastic process with countable state space E. Then $\mathcal{Z}$ is called a **semi–regenerative process** if there is an embedded Markov renewal chain $(\mathcal{X}, \mathcal{T})$ such that all T_n are stopping times for $\mathcal{Z}$, all X_n are deterministic functions of $(Z_u : u \leq T_n)$, and

$$\mathbb{P}(Z_{T_n + t_1} = j_1, \dots, Z_{T_n + t_k} = j_k | Z_u : u \leq T_n, X_n = i)$$
$$= \mathbb{P}(Z_{t_1} = j_1, \dots, Z_{t_k} = j_k | X_0 = i) \quad (7.5)$$

holds for all $n, k \in \mathbb{N}$, $i, j_1, \dots, j_k \in E$, and $t_1 < \dots < t_k \in \mathbb{R}_0^+$. This condition postulates that for any prediction of the process $(Z_u : u \geq T_n)$ all information of the past $(Z_u : u \leq T_n)$ is contained in the state X_n. We abbreviate

$$K_{ij}(t) := \mathbb{P}(T_1 > t, Z_t = j | X_0 = i)$$

for all $i, j \in E$ and $t \in \mathbb{R}_0^+$.

Theorem 7.14 *The transient distributions of a semi–regenerative process $\mathcal{Z}$ with embedded Markov renewal sequence $(\mathcal{X}, \mathcal{T})$ and initial distribution π are given by $\mathbb{P}(Z_t = j) = \sum_{i \in E} \pi_i P_{ij}(t)$ with*

$$P_{ij}(t) = \sum_{n=0}^{\infty} \sum_{k \in E} \int_0^t K_{kj}(t - u) \, dF_{ik}^{*n}(u)$$

for all $t > 0$ and $i, j \in E$.

Proof: This expression is obtained by conditioning upon the number n of Markov renewal points until time t and the state k which is observed at the last Markov renewal point before t.

$\square$

The following limit theorem is the main result of this chapter and will be applied to many classical queueing systems later on. We define the column vector $m = (m_i : i \in E)$ with

$$m_i := \mathbb{E}(T_1 | X_0 = i)$$

for all $i \in E$, and abbreviate $\nu m := \sum_{i \in E} \nu_i m_i$.

Theorem 7.15 *Let $\mathcal{Z}$ denote a semi–regenerative process with irreducible and positive recurrent embedded Markov chain $\mathcal{X}$. Denote the stationary distribution of $\mathcal{X}$ by ν and assume that $\nu m < \infty$. Then the limit*

$$\lim_{t \to \infty} \mathbb{P}(Z_t = j) = \frac{1}{\nu m} \sum_{k \in E} \nu_k \int_0^\infty K_{kj}(t) \, dt$$

holds for all $j \in E$ and is independent of the initial distribution.

Proof: Since $\mathcal{X}$ is positive recurrent and $\nu m < \infty$, the process $\mathcal{Z}$ is regenerative with regeneration times being the transition times to some state $i \in E$. Then theorem 7.2 and lemma 7.11 yield

$$\lim_{t \to \infty} \mathbb{P}(Z_t = j) = \frac{1}{\mu_{ii}} \int_0^\infty \mathbb{P}(Z_t = j, \tau_i > t | X_0 = i) \, dt$$

$$= \frac{\nu_i}{\nu m} \cdot \mathbb{E}_i \left(\int_0^{\tau_i} \mathbf{1}_{Z_t = j} \, dt \right)$$

with τ_i defined as in (7.4) and $\mathbb{E}_i$ denoting the conditional expectation given $X_0 = i$. Defining the stopping time $\sigma_i = \min\{n \in \mathbb{N} : X_n = i\}$, we can write $\tau_i = \sum_{n=1}^{\sigma_i} (T_n - T_{n-1})$. The semi–regenerative property (7.5) yields

$$\mathbb{E}_i \left(\int_{T_{n-1}}^{T_n} \mathbf{1}_{Z_t = j} \, dt \, \middle| \, Z_u : u \le T_{n-1}, X_{n-1} = k \right) = \mathbb{E}_k \left(\int_0^{T_1} \mathbf{1}_{Z_t = j} \, dt \right)$$

for all $n \in \mathbb{N}$ and $k \in E$. Hence we can write

$$\lim_{t \to \infty} \mathbb{P}(Z_t = j) = \frac{\nu_i}{\nu m} \cdot \mathbb{E}_i \left(\sum_{n=1}^{\sigma_i} \mathbb{E}_{X_{n-1}} \left(\int_0^{T_1} \mathbf{1}_{Z_t = j} \, dt \right) \right)$$

$$= \frac{\nu_i}{\nu m} \cdot \mathbb{E}_i \left(\sum_{n=1}^{\sigma_i} \sum_{k \in E} \mathbf{1}_{X_{n-1} = k} \cdot \mathbb{E}_k \left(\int_0^{T_1} \mathbf{1}_{Z_t = j} \, dt \right) \right)$$

$$= \frac{\nu_i}{\nu m} \cdot \sum_{k \in E} \mathbb{E}_i \left(\sum_{n=1}^{\sigma_i} \mathbf{1}_{X_{n-1} = k} \right) \cdot \mathbb{E}_k \left(\int_0^{T_1} \mathbf{1}_{Z_t = j} \, dt \right)$$

By theorems 2.24 and 2.27 we get

$$\mathbb{E}_i \left(\sum_{n=1}^{\sigma_i} 1_{X_{n-1}=k} \right) = \frac{\nu_k}{\nu_i}$$

whence the statement follows.

$\square$

Notes

Early papers on Markov renewal theory go back to Pyke [69, 70]. Classical textbooks on semi–Markov processes are Ross [74, 75] and Çinlar [25], the latter containing further an extensive presentation on semi–regenerative processes. The proof for theorem 7.15 is due to Asmussen [5]. For more advanced material on regenerative processes see Kalashnikov [44].

Exercise 7.1 A regenerative process is called a **delayed regenerative process** if the respective sequence $\mathcal{T} = (T_n : n \in \mathbb{N})$ of stopping times is a delayed renewal process. Prove theorem 7.2 for delayed regenerative processes.

Exercise 7.2 Prove theorem 7.7.

Exercise 7.3 Consider a machine that switches states between "on" and "off". First it is switched on for an amount X_1 of time, then it is off for an amount Y_1 of time, followed by an amount X_2 of time switched on, and so forth. Assume that the sequences $(X_n : n \in \mathbb{N})$ and $(Y_n : n \in \mathbb{N})$ are both iid with distribution functions F and G for X_1 and Y_1, respectively. Give a model for the state of the machine in terms of a semi–Markov process and show that for $F * G$ being not lattice and $\mathbb{E}(X_1 + Y_1) < \infty$ the long–run fraction π_{on} of time that the machine is switched on can be expressed as

$$\pi_{on} = \frac{\mathbb{E}(X_1)}{\mathbb{E}(X_1) + \mathbb{E}(Y_1)}$$

Such a process is called an **alternating renewal process**.

Exercise 7.4 For a positive recurrent Markov process with discrete state space E, derive an expression for the mean recurrence time to a state $i \in E$.

Chapter 8

SEMI–MARKOVIAN QUEUES

The term semi–Markovian queues signifies the class of queues that can be analyzed by means of an embedded semi–Markov process, i.e. by modelling the system process of the queue as a semi–regenerative process.

1. The GI/M/1 Queue

The first queue that shall serve as an example for an application of the semi–Markovian method is the **GI/M/1 queue**. This has an arrival stream which is a renewal process, i.e. the inter–arrival times are iid with some common distribution function A. There is one single server with exponential service time distribution. Its **intensity**, i.e. the parameter of the exponential distribution, shall be denoted by $\mu > 0$. The service displine is FCFS and the capacity of the waiting room is unbounded.

The system process $\mathcal{Q} = (Q_t : t \in \mathbb{R}_0^+)$ has state space $E = \mathbb{N}_0$, with Q_t indicating the number of users in the system (i.e. in service or waiting) at time t. It can be modelled by a Markov process only for the case of exponential inter–arrival times, i.e. for $A(t) = 1 - e^{-\lambda t}$ with some $\lambda > 0$, since only the exponential distribution is memoryless (see section 4). For general distribution functions A, we need to find another method of analysis.

One feature that clearly distinguishes this particular queueing system GI/M/1 is the independence of the arrival process from the rest of the system, which leads immediately to the construction of an embedded Markov renewal sequence at times of arrivals. This is possible since at times of arrival we know that the new inter–arrival time has just begun and because of the memoryless service time distribution we do not need to remember anything else than the number of users in the system. Thus an analysis of the system as a semi–regenerative process seems appropriate. That this is indeed successful will be shown in the following.

Define T_n as the time of the nth arrival and $X_n := Q_{T_n} - 1$ as the number of users in the system immediately before the nth arrival. Clearly, the T_n are stopping times and X_n is a deterministic function of Q_{T_n}. Assume that A is not lattice. Further we postulate $A(0) = 0$ and $\mathbb{E}(A) < \infty$. This implies in particular that $T_n \to \infty$ almost surely as n tends to infinity. The sequence $\mathcal{X} = (X_n : n \in \mathbb{N}_0)$ is a Markov chain since at times of arrivals the future of the system is determined only by the current number of users in the system, due to the memoryless property of the service times. The same property of the queue ensures the validity of equation (7.5) for the system process $\mathcal{Q}$.

Thus the system process $\mathcal{Q}$ is semi–regenerative with embedded Markov renewal chain $(\mathcal{X}, \mathcal{T})$, denoting $\mathcal{T} = (T_n : n \in \mathbb{N}_0)$ with $T_0 := 0$. For the characterizing matrix F of $(\mathcal{X}, \mathcal{T})$ we obtain

$$F_{ij}(x) = \mathbb{P}(T_{n+1} - T_n \leq x, X_{n+1} = j | X_n = i)$$

$$= \begin{cases} \int_0^x e^{-\mu t} \frac{(\mu t)^{i+1-j}}{(i+1-j)!} \, dA(t), & 1 \leq j \leq i+1 \\ 1 - \sum_{k=0}^i F_{ik}(x), & j = 0 \\ 0, & j > i+1 \end{cases}$$

for all $i, j \in E = \mathbb{N}_0$. The transition probability matrix $P = (p_{ij})_{i,j\in\mathbb{N}_0}$ of $\mathcal{X}$ is given by its entries

$$p_{ij} = \begin{cases} a_{i+1-j} := \int_0^\infty e^{-\mu t} \frac{(\mu t)^{i+1-j}}{(i+1-j)!} \, dA(t), & 1 \leq j \leq i+1 \\ b_i := 1 - \sum_{n=0}^i a_n, & j = 0 \\ 0, & j > i+1 \end{cases}$$

for all $i, j \in \mathbb{N}_0$. Here the first line describes the case that within an inter–arrival time exactly $i + 1 - j$ users are served such that immediately before the next arrival there are j users in the system, given that i users have been in the system immediately before the last arrival. The third line states that after one inter–arrival period there can only be an increase by one user in the system. The second line distributes the remaining probability mass to the only possible case left.

Clearly, $b_n = \sum_{k=n+1}^\infty a_k$ holds for all $n \in \mathbb{N}_0$. With the abbreviations a_n and b_n, the matrix P is strucured as

$$P = \begin{pmatrix} b_0 & a_0 & 0 & 0 & 0 & \dots \\ b_1 & a_1 & a_0 & 0 & 0 & \dots \\ b_2 & a_2 & a_1 & a_0 & 0 & \dots \\ \vdots & \vdots & \ddots & \ddots & \ddots & \ddots \end{pmatrix} \qquad (8.1)$$

Such a matrix is called **upper Hessenberg matrix** or **skip–free to the right**. It is characterized by the feature that above the first diagonal on top of the main diagonal the matrices contain only zeros.

The function $K(t)$ describing the behaviour of the system process between Mar-kov renewal points is given by $K_{ij}(t) = \mathbb{P}(T_1 > t, Q_t = j | X_0 = i)$ for $i, j \in \mathbb{N}_0$. Exploiting the independence of arrival process and service we obtain

$$
K_{ij}(t) = \begin{cases} (1 - A(t)) \cdot e^{-\mu t} \frac{(\mu t)^{i+1-j}}{(i+1-j)!}, & 1 \le j \le i+1 \\ (1 - A(t)) \cdot e^{-\mu t} \sum_{n=i+1}^{\infty} \frac{(\mu t)^n}{n!}, & j = 0 \\ 0, & j > i+1 \end{cases} \tag{8.2}
$$

for all $t > 0$, and $i, j \in \mathbb{N}_0$. The transient distributions of the system process can now be determined according to theorem 7.14.

In order to employ theorem 7.15 for the calculation of the asymptotic distribution of the system process, we need to determine the stationary distribution ν of $\mathcal{X}$, and the vector m of the mean time between Markov renewal points. The vector m is obtained in a straightforward manner as

$$
m_i = \mathbb{E}(T_1 | X_0 = i) = \mathbb{E}(A) \tag{8.3}
$$

independently of $i \in \mathbb{N}_0$, since the arrival process does not depend on the number of users in the system. Thus the vector m is constant.

The most difficult part is to determine the stationary distribution ν of the Markov chain $\mathcal{X}$. Since $a_0 > 0$ and $b_n > 0$ for all $n \in \mathbb{N}_0$, the transition matrix P is clearly irreducible. Hence there is at most one stationary distribution of $\mathcal{X}$. The stationary distribution ν of $\mathcal{X}$ can be determined by solving the following system of equations:

$$
\nu_0 = \sum_{n=0}^{\infty} \nu_n b_n \quad \text{and} \quad \nu_k = \sum_{n=k-1}^{\infty} \nu_n a_{n-k+1} \tag{8.4}
$$

With a geometric approach, i.e. assuming $\nu_n = (1 - \xi) \cdot \xi^n$ for all $n \in \mathbb{N}_0$ and some $0 < \xi < 1$, the equations for $k \ge 1$ can be transformed to

$$
(1 - \xi)\xi^k = (1 - \xi) \sum_{n=k-1}^{\infty} \xi^n a_{n-k+1} \quad \Leftrightarrow \quad \xi = \sum_{n=0}^{\infty} \xi^n a_n \tag{8.5}
$$

If some $0 < \xi < 1$ satisfies this equation, then

$$\nu_0 = \sum_{n=0}^{\infty} \nu_n b_n = (1-\xi) \sum_{n=0}^{\infty} \xi^n \sum_{k=n+1}^{\infty} a_k = (1-\xi) \sum_{k=1}^{\infty} \sum_{n=0}^{k-1} a_k \xi^n$$

$$= \sum_{k=1}^{\infty} a_k (1-\xi^k) = 1 - a_0 - \sum_{k=1}^{\infty} a_k \xi^k = 1 - a_0 - (\xi - a_0) = 1 - \xi$$

holds, too. This means that the approach

$$\nu_n = (1-\xi)\xi^n \qquad \text{for all } n \in \mathbb{N}_0 \tag{8.6}$$

would yield a stationary distribution for $\mathcal{X}$ if the number $0 < \xi < 1$ satisfying $\xi = \sum_{n=0}^{\infty} \xi^n a_n$ can be determined.

To this aim, we consider the power series $f(x) = \sum_{n=0}^{\infty} a_n x^n$ which is well–defined on the interval $[0, 1]$. Clearly, $f(1) = 1$ and $f(0) = a_0 > 0$. Since $a_n > 0$ for all $n \in \mathbb{N}_0$, we obtain $f''(x) > 0$ for all x which means that the function f is strictly convex on $[0, 1]$. A fix point $\xi = \sum_{n=0}^{\infty} \xi^n a_n$ geometrically signifies the first coordinate of an intersection between f and the identity function.

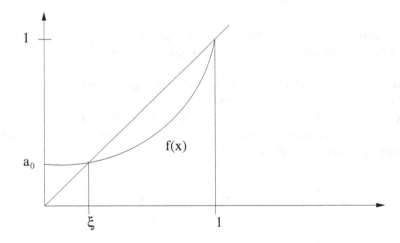

Figure 8.1. Fix point as intersection with diagonal

The above properties of f and the mean value theorem together imply that such a fix point ξ does exist if and only if the condition $f'(1) > 1$ is satisfied.

Because of

$$f'(1) = \sum_{n=1}^{\infty} n a_n = \int_0^{\infty} e^{-\mu t} \sum_{n=1}^{\infty} n \frac{(\mu t)^n}{n!} \, dA(t)$$

$$= \int_0^{\infty} e^{-\mu t} \sum_{n=1}^{\infty} \frac{(\mu t)^{n-1}}{(n-1)!} (\mu t) \, dA(t) = \mu \cdot \int_0^{\infty} t \, dA(t)$$

this condition translates to

$$\mu \cdot \mathbb{E}(A) > 1 \qquad \Leftrightarrow \qquad \mathbb{E}(A) > \frac{1}{\mu} \tag{8.7}$$

which simply means that the mean inter–arrival time is strictly greater than the mean service time.

Remark 8.1 One way to calculate the number ξ is the following: We start with $\xi_0 := 0$ and iterate by $\xi_{n+1} := f(\xi_n)$ for all $n \in \mathbb{N}_0$. Then $\xi = \lim_{n \to \infty} \xi_n$. In order to prove this we first observe that the sequence $(\xi_n : n \in \mathbb{N}_0)$ is strictly increasing. This follows by induction, as $\xi_1 = a_0 > 0 = \xi_0$ and

$$\xi_{n+1} = \sum_{k=0}^{\infty} a_k \xi_n^k > \sum_{k=0}^{\infty} a_k \xi_{n-1}^k = \xi_n$$

by induction hypothesis, since all a_n are strictly positive. Again by induction we obtain $\xi_n < 1$ for all $n \in \mathbb{N}_0$, as $\xi_0 < 1$ and

$$\xi_n = \sum_{k=0}^{\infty} a_k \xi_{n-1}^k < \sum_{k=0}^{\infty} a_k = 1$$

for all $n \in \mathbb{N}$. Hence the sequence $(\xi_n : n \in \mathbb{N}_0)$ converges as it is increasing and bounded. Now for the limit ξ_{∞} we obtain

$$\xi_{\infty} = \lim_{n \to \infty} \xi_n = \lim_{n \to \infty} \xi_{n+1} = \lim_{n \to \infty} \sum_{k=0}^{\infty} a_k \xi_n^k = \sum_{k=0}^{\infty} a_k \left(\lim_{n \to \infty} \xi_n \right)^k$$

$$= \sum_{k=0}^{\infty} a_k \xi_{\infty}^k$$

which means that ξ_{∞} is a fix point for the function f. This is strictly increasing, as $f'(x) > 0$ for all $x > 0$ due to the positivity of all a_n. Hence

$$f(x) < f(\xi) = \xi \quad \text{for all} \quad x < \xi$$

Since the sequence $(\xi_n : n \in \mathbb{N}_0)$ starts with $\xi_0 = 0$, this means that $\xi_n < \xi$ for all $n \in \mathbb{N}_0$. Hence $\xi_{\infty} = \xi$, since ξ is the only fix point smaller than one.

Now theorem 7.15 can be applied with the values for ν, m and K as determined above. This yields for the asymptotic distribution of the GI/M/1 queue the following results (see exercises):

$$\pi_j := \lim_{t \to \infty} \mathbb{P}(Q_t = j) = \frac{(1 - \xi)\xi^{j-1}}{\mathbb{E}(A)} \int_0^\infty (1 - A(t))e^{-\mu t(1-\xi)} \, dt \quad (8.8)$$

for all $j \geq 1$ and

$$\pi_0 := \lim_{t \to \infty} \mathbb{P}(Q_t = 0) = 1 - \frac{1}{\mathbb{E}(A)} \int_0^\infty (1 - A(t))e^{-\mu t(1-\xi)} \, dt \quad (8.9)$$

Because of

$$\xi = \sum_{n=0}^\infty a_n \xi^n = \int_0^\infty e^{-\mu t} \sum_{n=0}^\infty \frac{(\mu t \xi)^n}{n!} \, dA(t) = \int_0^\infty e^{-\mu t(1-\xi)} \, dA(t)$$

and integration by parts

$$\int_0^\infty A(t)e^{-\mu t(1-\xi)} \, dt = \frac{-1}{\mu(1-\xi)} \left[A(t)e^{-\mu t(1-\xi)} \right]_0^\infty$$
$$- \frac{-1}{\mu(1-\xi)} \int_0^\infty e^{-\mu t(1-\xi)} \, dA(t)$$
$$= \frac{\xi}{\mu(1-\xi)}$$

we obtain

$$\int_0^\infty (1 - A(t))e^{-\mu t(1-\xi)} \, dt = \int_0^\infty e^{-\mu t(1-\xi)} \, dt - \int_0^\infty A(t)e^{-\mu t(1-\xi)} \, dt$$
$$= \frac{1}{\mu(1-\xi)} - \frac{\xi}{\mu(1-\xi)} = \frac{1}{\mu} \quad (8.10)$$

Hence the asymptotic distributions are given by

$$\pi_0 = 1 - \rho \quad \text{and} \quad \pi_j = (1 - \xi)\xi^{j-1}\rho$$

for $j \geq 1$, with $\rho := (\mu \cdot \mathbb{E}(A))^{-1}$. The condition (8.7), which assures positive recurrence of the chain $\mathcal{X}$, is equivalent to the **stability condition** $\rho < 1$. This guarantees the existence of an asymptotic distribution of the system process. The value ρ is called the **load** of the queueing system.

If $\rho \geq 1$, which means that the mean service time is not smaller than the mean inter–arrrival time, we expect the queue to be unstable. For the system process Q and already for its embedded Markov chain $\mathcal{X}$ we would in this case expect that no asymptotic resp. stationary distributions exist. This will be shown in the remainder of this section.

To this aim we will use the concept of a subinvariant measure, which is defined as follows: Let E denote a countable space and P an irreducible stochastic matrix with state space E. A measure μ on E is called **subinvariant** for P if

$$\mu_j \geq \sum_{i \in E} \mu_i p_{ij}$$

holds for all $j \in E$. If there is an equality in the above relation, then μ is called **invariant** for P.

Let $\mathcal{X}$ denote a Markov chain with transition matrix P. Define the so–called **taboo probabilities** in n steps by

$$_T P^n(i, j) := \mathbb{P}(X_n = j, X_k \notin T \, \forall \, 0 < k < n | X_0 = i) \qquad (8.11)$$

for all $i, j \in E$ and the taboo set $T \subset E$. If $T = \{\alpha\}$ has only one element, we will write α instead of $\{\alpha\}$ as an index. Now we can show the following important result for irreducible stochastic matrices:

Theorem 8.2 *Let $\mathcal{X}$ denote an irreducible Markov chain with transition matrix P and countable state space E. Choose any state $\alpha \in E$. Then the measure μ^α defined by*

$$\mu_j^\alpha := \sum_{n=1}^{\infty} {}_\alpha P^n(\alpha, j)$$

*is the **minimal subinvariant measure** for P in the sense that for any other subinvariant measure μ with $\mu_\alpha = 1$ the relation $\mu_j \geq \mu_j^\alpha$ holds for all $j \in E$. Further μ^α is invariant for P if and only if P is recurrent. In this case, μ^α is the only subinvariant measure for P with $\mu_\alpha^\alpha = 1$.*

Proof: By definition of μ^α, we obtain

$$\sum_{i \in E} \mu_i^\alpha p_{ij} = \mu_\alpha^\alpha p_{\alpha,j} + \sum_{i \neq \alpha} \sum_{n=1}^{\infty} {}_\alpha P^n(\alpha, i) p_{ij}$$

$$\leq {}_\alpha P^1(\alpha, j) + \sum_{n=2}^{\infty} {}_\alpha P^n(\alpha, j)$$

$$= \mu_j^\alpha$$

where the inequality comes from the bound $\mu_\alpha^\alpha = \mathbb{P}(\tau_\alpha < \infty | X_0 = \alpha) \leq 1$, with τ_α denoting the first return time to state α. Thus μ^α is subinvariant and invariant if and only if $\mathbb{P}(\tau_\alpha < \infty | X_0 = \alpha) = 1$, i.e. if $\mathcal{X}$ is recurrent.

Let μ denote any other subinvariant measure for P with $\mu_\alpha = 1$. We will show by induction on n that $\mu_j \geq \sum_{k=1}^n {_\alpha P^k}(\alpha, j)$ holds for all $n \geq 1$. First, subinvariance of μ yields

$$\mu_j \geq \sum_{i \in E} \mu_i p_{ij} \geq \mu_\alpha p_{\alpha,j} = p_{\alpha,j} = {_\alpha P^1}(\alpha, j)$$

for all $j \in E$. The induction step follows from

$$\mu_j \geq \mu_\alpha p_{\alpha,j} + \sum_{i \neq \alpha} \mu_i p_{ij} \geq p_{\alpha,j} + \sum_{i \neq \alpha} \left(\sum_{k=1}^n {_\alpha P^k}(\alpha, i) \right) p_{ij}$$

$$= \sum_{k=1}^{n+1} {_\alpha P^k}(\alpha, j)$$

where the first inequality is due to the subinvariance of μ and the second one follows from the induction hypothesis. The minimality of μ^α follows in the limit $n \to \infty$.

Assume that $\mathcal{X}$ is recurrent which implies invariance of μ^α and $\mu_\alpha^\alpha = 1$. Let μ denote another subinvariant measure for P and assume $\mu_\alpha = 1$. If $\mu \neq \mu^\alpha$, then there is a state $j \in E$ and a number $n \in \mathbb{N}$ such that $\mu_j > \mu_j^\alpha$ and $P^n(j, \alpha) > 0$, due to the minimality of μ^α and irreducibility of P. Then we obtain

$$1 = \mu_\alpha \geq \sum_{i \in E} \mu_i \cdot P^n(i, \alpha) > \sum_{i \in E} \mu_i^\alpha \cdot P^n(i, \alpha) = \mu_\alpha^\alpha = 1$$

which is a contradiction. Hence there is no other subinvariant measure μ with $\mu_\alpha = 1$.
$\square$

Now we will apply this result to the embedded Markov chain $\mathcal{X}$ immediately before arrival times of a GI/M/1 queue. Define the sets $[k] := \{0, \ldots, k\}$ for all $k \in \mathbb{N}_0$. Because of the upper Hessenberg structure of the transition matrix P, we obtain

$$_0 P^n(0, k+1) = \sum_{l=1}^{n-1} {_0 P^l}(0, k) \cdot {_{[k]} P^{n-l}}(k, k+1) \tag{8.12}$$

for all $k \in \mathbb{N}$ by conditioning upon the time of the last visit to state k. The self–similarity of P yields $_{[k]} P^l(k, k+1) = {_0 P^l}(0, 1)$. Summing the equations in

(8.12) for all $n \geq 1$ yields

$$\sum_{n=1}^{\infty} {}_0 P^n(0, k+1) = \sum_{n=1}^{\infty} \sum_{l=1}^{n-1} {}_0 P^l(0, k) \cdot {}_{[k]} P^{n-l}(k, k+1)$$

$$= \sum_{l=1}^{\infty} {}_0 P^l(0, k) \sum_{n=l+1}^{\infty} {}_{[k]} P^{n-l}(k, k+1)$$

$$= \left(\sum_{n=1}^{\infty} {}_0 P^n(0, k) \right) \cdot \left(\sum_{n=1}^{\infty} {}_0 P^n(0, 1) \right)$$

Setting $\alpha = 0$, we obtain for the minimal subinvariant measure

$$\mu_{k+1}^0 = \mu_k^0 \cdot \mu_1^0 \qquad \Leftrightarrow \qquad \mu_k^0 = \mu_0^0 \cdot \xi^k$$

with $\xi = \mu_1^0$. Thus any invariant measure for P must have a geometric form. Because of (8.5), the parameter of this geometric distribution is given by the fix point $\xi = \sum_{n=1}^{\infty} a_n \xi^n$. If there is no solution $0 < \xi < 1$ for this equation, then there cannot exist a finite invariant measure and hence no stationary distribution.

2. The M/G/1 Queue

The classical counterpart of the GI/M/1 queue is the M/G/1 queue. It admits a similar method of analysis. The **M/G/1 queue** has a Poisson arrival process and general iid service time distributions. One server works according to the FCFS discipline. The waiting room is infinite such that there are no lost users. Denote the arrival intensity, which is the parameter of the Poisson arrival process, by $\lambda > 0$. Further denote the distribution function of the service time by B.

In the GI/M/1 queue, the exponential service time appears to be a rather special assumption, since telephone calls, demands on server capacity etc. are certainly not memoryless. This peculiarity translates to memoryless inter–arrival times for the M/G/1 counterpart. However, the following theorem 8.3 shows that this property follows from a few assumptions which can often be observed in classical fields of application, such as the classical (voice only) telephone network.

A pure jump process $\mathcal{A} = (A_t : t \in \mathbb{R}_0^+)$ with state space $\mathbb{N}_0$ and the property that $A_t \geq A_s$ for all $t > s$ is called a **counting process** or an **arrival process**. For an arrival process in a classical telephone network we may assume the following properties. There is a large population of possible users which all act independently from each others. Hence only a finite number of users can decide to use the system within a finite time interval. Their decision to enter the system, i.e. to use the system capacity, is homogeneous in time.

Theorem 8.3 *Let $\mathcal{A}$ denote an arrival process with the following properties:*
(a) Only a finite number of arrivals occur within a finite time interval.
(b) The process $\mathcal{A}$ has indendent increments, i.e. the random variables $A_t - A_s$ and $A_v - A_u$ are independent for all $s < t \leq u < v$.
(c) The process $\mathcal{A}$ has stationary increments, i.e. the distribution of $A_t - A_s$ for any $s < t$ depends only on the distance $t - s$.
(d) There are only single arrivals, i.e. $\lim_{h \to 0}(A_{t+h} - A_t) \leq 1$ almost certainly for all $t \geq 0$.
Then $\mathcal{A}$ is a homogeneous Poisson process.

Proof: Let $\Phi(t)$ denote the probability that no arrivals occur in the interval $[0, t[$, i.e. define

$$\Phi(t) := \mathbb{P}(A_t - A_0 = 0)$$

for all $t \geq 0$. By definition, $0 \leq \Phi(t) \leq 1$ for all $t \geq 0$. Because of assumption (a), the function Φ does not vanish identically, i.e. $\Phi \neq 0$. Assumptions (b) and (c) imply

$$\Phi(s + t) = \Phi(s) \cdot \Phi(t) \tag{8.13}$$

for all $s, t > 0$. If there were a time $t_0 > 0$ with $\Phi(t_0) = 0$, then property (8.13) would yield

$$0 = \Phi(t_0) = (\Phi(t_0/2))^2 = (\Phi(t_0/4))^4 = \ldots$$

such that Φ would vanish arbitrarily near $t = 0$ and therefore $\Phi = 0$ since Φ is monotone non–increasing. Hence $\Phi(t) > 0$ for all $t \geq 0$. The only monotone non–vanishing solution to (8.13) is given by

$$\Phi(t) = e^{-\lambda t}$$

for some value $\lambda \geq 0$. By assumptions (b) and (c), this means that the time until the first arrival is distributed exponentially with parameter λ, where λ is independent of the number of arrivals that have already occurred. By assumption (d) there is only one arrival at a time. These two conditions are exactly the defining properties of the Poisson process as given in example 6.5. The case $\lambda = 0$ corresponds to the singular case where no arrivals occur ever.
$\square$

Again, the system process $\mathcal{Q} = (Q_t : t \in \mathbb{R}_0^+)$ has state space $E = \mathbb{N}_0$ and Q_t denotes the number of users in the system at time t. Similar to the GI/M/1 queue we can construct an embedded Markov chain by considering the number of users in the system immediately after service completions. At these time instances we know that the current service (if there is one) has just begun and need only to remember the number of users in the system in order to predict this number immediately after the next service completion.

Define $T_0 := 0$ and T_n as the time of the nth service completion. Further define $\mathcal{T} := (T_n : n \in \mathbb{N}_0)$. Let $X_n := Q_{T_n}$ for all $n \in \mathbb{N}_0$ and assume that at the time origin there are no users in the system. The T_n are stopping times for $\mathcal{Q}$ and by definition X_n is a deterministic function of Q_{T_n}. Assume that $0 < \mathbb{E}(B) < \infty$. This implies $T_n \to \infty$ for $n \to \infty$. As shown above, the chain $\mathcal{X} = (X_n : n \in \mathbb{N}_0)$ is a Markov chain due to the fact that at service completion times there is either no new service (if the system is empty) or a new service has just begun. The same property of the queue yields condition (7.5). Hence $\mathcal{Q}$ is a semi–regenerative process with embedded Markov renewal chain $(\mathcal{X}, \mathcal{T})$.

The transition matrix P of $\mathcal{X}$ is given by its entries

$$
p_{ij} = \begin{cases} a_k := \int_0^\infty e^{-\lambda t} \frac{(\lambda t)^k}{k!} \, dB(t), & j = i - 1 + k \\ 0, & j < i - 1 \end{cases}
$$

for $i \geq 1$, and $p_{0,j} = a_j$ for all $j \in \mathbb{N}_0$. The first line above describes the fact that during one service time there need to be $k = j - i + 1$ arrivals in order to observe j users in the system after the next service completion if there were i users after the last service completion. The second line simply states that within one service time not more than one user can leave the system. The entries $p_{0,j}$ are explained by the fact that if the system is empty after a service completion, then the next service starts only after an arrival has occurred and hence $p_{0,j} = p_{1,j}$. The matrix P is structured as

$$
P = \begin{pmatrix} a_0 & a_1 & a_2 & a_3 & \cdots \\ a_0 & a_1 & a_2 & a_3 & \cdots \\ 0 & a_0 & a_1 & a_2 & \cdots \\ \vdots & \ddots & \ddots & \ddots & \ddots \end{pmatrix}
$$

Such a matrix is called **lower Hessenberg matrix** or **skip–free to the left**. The Markov renewal chain $(\mathcal{X}, \mathcal{T})$ is characterized by $F_{ij}(t) = p_{ij} \cdot G_{ij}(t)$, with

$$
G_{ij}(t) = \mathbb{P}(T_{n+1} - T_n \leq t | X_n = i, X_{n+1} = j)
$$
$$
= \begin{cases} B(t), & i > 0 \\ \int_0^t e^{-\lambda u} \lambda B(t - u) \, du, & i = 0 \end{cases}
$$

independently of $j \in \mathbb{N}_0$. The values $K_{ij}(t) = \mathbb{P}(T_1 > t, Q_t = j | X_0 = i)$ are given by

$$K_{ij}(t) = \begin{cases} e^{-\lambda t}, & i = j = 0 \\ \int_0^t e^{-\lambda u} \lambda e^{-\lambda(t-u)} \frac{(\lambda(t-u))^{j-1}}{(j-1)!} (1 - B(t-u)) \, du, & i = 0, j > 0 \\ (1 - B(t)) \cdot e^{-\lambda t} \frac{(\lambda t)^{j-i}}{(j-i)!}, & 0 < i \le j \end{cases}$$

for all $t > 0$.

Since all a_n are strictly positive, the matrix P is clearly irreducible. Assume that $\sum_{n=1}^\infty n a_n < 1$ and set $\varepsilon := 1 - \sum_{n=1}^\infty n a_n$. Further define the function $f(n) := n$ for all $n \in \mathbb{N}_0$. Because of

$$\sum_{j=0}^\infty p_{ij} f(j) = \sum_{j=i-1}^\infty a_{j-i+1} j = \sum_{j=i-1}^\infty a_{j-i+1}(j - i + 1) + i - 1$$

$$= \sum_{n=1}^\infty n a_n - 1 + f(i) \le f(i) - \varepsilon$$

for all $i \ge 1$, the function f satisfies Foster's criterion (see theorem 2.33) and hence $\mathcal{X}$ is positive recurrent if the condition $\sum_{n=1}^\infty n a_n < 1$ holds. Since

$$\sum_{n=1}^\infty n a_n = \sum_{n=1}^\infty n \int_0^\infty e^{-\lambda t} \frac{(\lambda t)^n}{n!} \, dB(t) = \int_0^\infty \sum_{n=1}^\infty e^{-\lambda t} \frac{(\lambda t)^{n-1}}{(n-1)!} (\lambda t) \, dB(t)$$

$$= \lambda \cdot \mathbb{E}(B)$$

the above condition is equivalent to the **stability condition**

$$\rho := \lambda \cdot \mathbb{E}(B) < 1 \qquad \Leftrightarrow \qquad \mathbb{E}(B) < \frac{1}{\lambda} \qquad (8.14)$$

which simply states that the mean service time be strictly smaller than the mean inter–arrival time.

In order to obtain the stationary distribution ν for $\mathcal{X}$, we need to solve the equation system

$$\nu_0 = \nu_0 a_0 + \nu_1 a_0$$
$$\nu_1 = \nu_0 a_1 + \nu_1 a_1 + \nu_2 a_0$$
$$\nu_2 = \nu_0 a_2 + \nu_1 a_2 + \nu_2 a_1 + \nu_3 a_0$$

$\cdots$

For each line k, adding the first k equations yields the recursive scheme

$$a_0\nu_1 = \nu_0 r_0$$
$$a_0\nu_2 = \nu_0 r_1 + \nu_1 r_1 \tag{8.15}$$
$$a_0\nu_3 = \nu_0 r_2 + \nu_1 r_2 + \nu_2 r_1$$

$$\ldots$$

with the abbreviations $r_k := 1 - \sum_{n=0}^{k} a_n = \sum_{n=k+1}^{\infty} a_n$. Define further $r := \sum_{n=0}^{\infty} r_n$ and note that $r = \sum_{n=1}^{\infty} n a_n = \rho < 1$ and $a_0 = 1 - r_0$. Adding all these equations yields

$$(1 - r_0) \sum_{n=1}^{\infty} \nu_n = \nu_0 r + \sum_{n=1}^{\infty} \nu_n(r - r_0) \tag{8.16}$$

which implies

$$\sum_{n=1}^{\infty} \nu_n = \nu_0 \frac{r}{1 - r} \qquad \Longleftrightarrow \qquad 1 = \nu_0 \frac{r + (1 - r)}{1 - r} = \frac{\nu_0}{1 - \rho}$$

Hence the stationary distribution of $\mathcal{X}$ is given by $\nu_0 = 1 - \rho$ and the recursive scheme above.

Theorem 8.4 *If the stability condition (8.14) holds, then the asymptotic distribution of the M/G/1 queue is given by*

$$\pi_j := \lim_{t \to \infty} \mathbb{P}(Q_t = j) = \nu_j$$

for all $j \in \mathbb{N}_0$.

Proof: By theorem 7.15 and the fact that $K_{ij}(t) = 0$ for $i > j$ we obtain

$$\pi_j = \frac{1}{\nu m} \sum_{i=0}^{j} \nu_i \int_0^{\infty} K_{ij}(t)\, dt$$

for all $j \in E$. For the mean renewal interval we get

$$\nu m = \nu_0 \left(\frac{1}{\lambda} + \mathbb{E}(B) \right) + \sum_{n=1}^{\infty} \nu_n \mathbb{E}(B) = \frac{1 - \rho}{\lambda} + \mathbb{E}(B) = \frac{1}{\lambda}$$

Thus it remains to prove that

$$\pi_j = \lambda \sum_{i=0}^{j} \nu_i \int_0^{\infty} K_{ij}(t)\, dt = \nu_j$$

holds for all $j \in E$. Abbreviate $\bar{B}(t) := 1 - B(t)$ for all $t \geq 0$. First we observe for $j > 0$

$$\int_0^\infty K_{0,j}(t)\, dt = \int_0^\infty \int_0^t e^{-\lambda u} \lambda e^{-\lambda(t-u)} \frac{(\lambda(t-u))^{j-1}}{(j-1)!} \bar{B}(t-u)\, du\, dt$$

$$= \int_0^\infty e^{-\lambda s} \lambda\, ds \cdot \int_0^\infty e^{-\lambda t} \frac{(\lambda t)^{j-1}}{(j-1)!} \bar{B}(t)\, dt$$

$$= \int_0^\infty K_{1,j}(t)\, dt$$

For $1 \leq i \leq j$ it can be shown (see exercises) that

$$\int_0^\infty K_{ij}(t)\, dt = \int_0^\infty e^{-\lambda t} \frac{(\lambda t)^{j-i}}{(j-i)!} \bar{B}(t)\, dt$$

$$= \frac{1}{\lambda} \int_0^\infty \sum_{k=j-i+1}^\infty e^{-\lambda t} \frac{(\lambda t)^k}{k!}\, dB(t) \qquad (8.17)$$

$$= \frac{r_{j-i}}{\lambda}$$

the last equality by definition of $(a_n)_{n \in \mathbb{N}_0}$ and r_{j-i}. Hence we obtain for $j \geq 1$

$$\pi_j = \nu_0 r_{j-1} + \sum_{i=1}^j \nu_i r_{j-i} = \nu_j$$

according to (8.15). As π and ν are probability distributions and we have shown that $\pi_j = \nu_j$ for all $j \geq 1$, we can infer $\pi_0 = \nu_0$ as well.
□

As we have done for the GI/M/1 queue, we want to show now that in case of $\rho \geq 1$ there is no stationary distribution of the embedded Markov chain $\mathcal{X}$ at times of service completion. We will prove this by contradiction. Assume that $\rho \geq 1$ and ν be a stationary distribution. Then the recursion scheme and in particular equation (8.16) holds, regardless of the value for $\rho = r$.

For the case $r = 1$ we obtain from this equation $0 = \nu_0$, which implies further $\nu_n = 0$ for all $n \geq 1$ because of P being irreducible. This is in contradiction to the assumption that ν be a distribution. In the case $r > 1$, we get from equation (8.16) that $\sum_{n=1}^\infty \nu_n = r > 1$, again a contradiction to the assumption that ν be a distribution.

3.   The GI/M/m Queue

The last example for semi–Markovian queues is a multi–server queue with exponential servers. Denote all the parameters as for the GI/M/1 queue, i.e.

the inter–arrival times are iid with distribution function A, and every server has an exponential distribution with parameter $\mu > 0$. Of course, instead of one server as in section 1, we now have m servers. The service displine is FCFS and the capacity of the waiting room is unbounded.

Regardless of the number of servers, all of them are memoryless. Hence the times of arrivals lead to an embedded Markov renewal chain, just as in the case $m = 1$. The system process $\mathcal{Q} = (Q_t : t \in \mathbb{R}_0^+)$ has state space $E = \mathbb{N}_0$, with Q_t indicating the number of users in the system (i.e. in service or waiting) at time t. Define T_n as the time of the nth arrival and $X_n := Q_{T_n} - 1$ as the number of users in the system immediately before the nth arrival.

Clearly, the T_n are stopping times and X_n is a deterministic function of Q_{T_n}. Again we assume that A is not lattice, and $0 < \mathbb{E}(A) < \infty$. This implies in particular that $T_n \to \infty$ almost surely as n tends to infinity. The sequence $\mathcal{X} = (X_n : n \in \mathbb{N}_0)$ is a Markov chain since at times of arrivals the future of the system is determined only by the current number of users in the system, due to the memoryless property of the service times. The same property of the queue ensures the validity of equation (7.5) for the system process $\mathcal{Q}$. Thus the system process $\mathcal{Q}$ is semi–regenerative with embedded Markov renewal chain $(\mathcal{X}, \mathcal{T})$.

The transition probabilities p_{ij} of $\mathcal{X}$ are derived by the following considerations. Clearly, there may be only one arrival in any interval $]T_n, T_{n+1}]$. Hence $p_{ij} = 0$ for $j > i + 1$.

Now consider the case $j \leq i + 1 \leq m$. This means that during one inter–arrival period no user is waiting and all are served with exponential intensity μ. Given that the inter–arrival period has length $t > 0$, the probability for any user to complete service is $1 - e^{-\mu t}$. A transition from state i to state j for the Markov chain $\mathcal{X}$ means that $i + 1 - j$ out of $i + 1$ users are served during one inter–arrival period. There are $\binom{i+1}{i+1-j} = \binom{i+1}{j}$ combinations to choose which users complete their service and which do not. Hence for $j \leq i + 1 \leq m$ we obtain

$$p_{ij} = \int_0^\infty \binom{i+1}{j} (1 - e^{-\mu t})^{i+1-j} (e^{-\mu t})^j \, dA(t) \qquad (8.18)$$

by conditioning on the length t of the inter–arrival period. Note that the integrand is the binomial distribution with $i + 1$ degrees of freedom and parameter $e^{-\mu t}$ evaluated at $0 \leq j \leq i + 1$.

The third case is $m \leq j \leq i + 1$. Here, all m servers are busy during the complete inter–arrival period. Then the number of users served in time t is distributed like a Poisson process (see example 6.5) with intensity $m \cdot \mu$ evaluated at time t. Conditioning on the length of the inter–arrival period, we obtain

$$p_{ij} = \int_0^\infty \frac{(m\mu \cdot t)^{i+1-j}}{(i+1-j)!} e^{-m\mu \cdot t} \, dA(t) \tag{8.19}$$

for $m \leq j \leq i+1$. Note that this expression coincides with formula (8.18) for $m = j = i+1$. Furthermore it depends only on the difference $i+1-j$, but not on the values of i, j themselves.

The last case to consider is $j < m < i+1$. In this situation there are $i+1-m$ users waiting in the queue at the beginning of the inter–arrival period, while $m-j$ servers are idle at the end of it. This is a mixture between the second and the third case. First the regime of the latter governs, until there are no waiting users any more (i.e. the queue has emptied). Then the former case applies. Thus we condition first on the length t of the inter–arrival period and then on the time $u < t$ to empty the queue. The time to empty the queue has an Erlang distribution $E_{i+1-m}^{m\mu}$ with $i+1-m$ stages and intensity $m\mu$. After that the number of served users has a binomial distribution B_m^p with m degrees of freedom and parameter $p := e^{-\mu(t-u)}$. This leads to an expression

$$p_{ij} = \int_0^\infty \int_0^t B_m^p(m-j) \, dE_{i+1-m}^{m\mu}(u) \, dA(t)$$
$$= \int_0^\infty \binom{m}{j} e^{-j\mu t} \int_0^t \frac{(m\mu u)^{i-m}}{(i-m)!} (e^{-\mu u} - e^{-\mu t})^{m-j} m\mu \, du \, dA(t) \tag{8.20}$$

which the reader may verify as an exercise.

Collecting these results, we can sketch the structure of the transition matrix as

$$P = \left(\begin{array}{cccc|cccc}
p_{00} & p_{01} & & & & & & \\
p_{10} & p_{11} & p_{12} & & & & & \\
\vdots & & \ddots & & & & & \\
p_{m-2,0} & \cdots & & p_{m-2,m-1} & & & & \\
p_{m-1,0} & \cdots & & p_{m-1,m-1} & \beta_0 & & & \\
\hline
p_{m,0} & \cdots & & p_{m,m-1} & \beta_1 & \beta_0 & & \\
\vdots & & & \vdots & \vdots & & \ddots & \\
p_{m+n,0} & \cdots & & p_{m+n,m-1} & \beta_{n+1} & \beta_n & \cdots & \beta_0 \\
\vdots & & & \vdots & \vdots & \vdots & & \ddots
\end{array} \right)$$

abbreviating $\beta_k := p_{i,i+1-k}$ in case (8.19). The non–specified entries in P correspond to the case $j > i+1$ and hence are zero. The matrix P has an upper Hessenberg form like the respective transition matrix for the GI/M/1 queue. The most important part of it is the lower right–hand part containing the entries β_n. The other parts are boundary conditions.

In order to determine the stationary distribution of $\mathcal{X}$, we first consider the partition $E = F \cup F^c$ with $F = \{0, \ldots, m-2\}$. We then obtain for the transition matrix P' of the Markov chain restricted to F^c the form

$$P' = \begin{pmatrix} r_0 & \beta_0 & & & \\ r_1 & \beta_1 & \beta_0 & & \\ r_2 & \beta_2 & \beta_1 & \beta_0 & \\ \vdots & \vdots & \ddots & \ddots & \ddots \end{pmatrix}$$

with $r_n = 1 - \sum_{k=0}^{n} \beta_k$. This has the same form as the transition matrix (8.1) for the GI/M/1 queue. Define

$$\rho = \frac{1}{m\mu \cdot \mathbb{E}(A)}$$

The arguments in section 1 all apply to the matrix P' with ρ defined as above. In the case $\rho < 1$ we obtain a stationary distribution $\nu' = \nu' P'$ given by

$$\nu'_n = (1 - \xi)\xi^n \tag{8.21}$$

with $\xi = \sum_{n=0}^{\infty} \xi^n \beta_n$. For the case $\rho \geq 1$ it follows that P' (and hence P) is not positive recurrent. In the following we assume $\rho < 1$.

By theorem 2.30 there is a unique extension of ν' to a stationary measure for P, denoted by $\nu'' = \nu'' P$. For this we have $\nu''_n = \nu'_{n-m+1}$ for all $n \geq m-1$ and

$$\nu''_{n+1} = \sum_{k=0}^{\infty} \nu''_k p_{k,n+1} = \sum_{k=n}^{\infty} \nu''_k p_{k,n+1}$$

for all $n = 0, \ldots, m-2$, which leads to

$$\nu''_n = \frac{1}{p_{n,n+1}} \left(\nu''_{n+1} \cdot (1 - p_{n+1,n+1}) - \sum_{k=n+2}^{\infty} \nu''_k p_{k,n+1} \right) \tag{8.22}$$

Thus ν''_{m-2} and iteratively $\nu''_{m-3}, \ldots, \nu''_0$ can be determined by formula (8.22). The stationary distribution $\nu = \nu P$ for $\mathcal{X}$ is then given as $\nu = c \cdot \nu''$ for a constant $c > 0$. This is determined by

$$c^{-1} = \sum_{n=0}^{\infty} \nu''_n = \sum_{n=0}^{m-2} \nu''_n + 1 \tag{8.23}$$

Altogether, formulae (8.21), (8.22), and (8.23) yield the stationary distribution $\nu = \nu P$ for $\mathcal{X}$.

The asymptotic distribution π of the queueing process $\mathcal{Q}$ can now be obtained by theorem 7.15. For $n \geq m$ we obtain

$$
\begin{aligned}
\pi_n &= \frac{c}{\mathbb{E}(A)} \sum_{k=n-1}^{\infty} (1-\xi)\xi^{k-(m-1)} \int_0^{\infty} (1-A(t))e^{-m\mu t} \frac{(m\mu t)^{k+1-n}}{(k+1-n)!} dt \\
&= \frac{c \cdot (1-\xi)}{\mathbb{E}(A)} \int_0^{\infty} (1-A(t))e^{-m\mu t}\xi^{n-m} \sum_{k=n-1}^{\infty} \frac{(m\mu t \cdot \xi)^{k+1-n}}{(k+1-n)!} dt \\
&= \frac{c \cdot (1-\xi)}{\mathbb{E}(A)} \xi^{n-m} \int_0^{\infty} (1-A(t))e^{-m\mu t \cdot (1-\xi)} dt
\end{aligned}
$$

Due to (8.10) the integral above equals $(m\mu)^{-1}$ and thus we get to

$$
\pi_n = \rho \cdot c \cdot (1-\xi)\xi^{n-m} \tag{8.24}
$$

for $n \geq m$. The asymptotic probability that all servers are busy is given by

$$
\sum_{n=m}^{\infty} \pi_n = \rho \cdot c
$$

Hence the conditional asymptotic probability that there are k users in the queue (i.e. $k + m$ users in the system) given that all servers are busy equals

$$
\frac{\pi_{m+k}}{\sum_{n=m}^{\infty} \pi_n} = (1-\xi)\xi^k
$$

which means that the conditional asymptotic queue length distribution given that all servers are busy is geometric.

Concerning the asymptotic probabilities π_n with $n = 1, \ldots, m-1$, we employ the **rate conservation law**, for which we first give an intuitive explanation. The arrival process of the GI/M/m queue is a renewal process with renewal intervals distributed by A. Blackwell's theorem 6.14 states that asymptotically an arrival occurs with constant rate $(\mathbb{E}(A))^{-1}$. Given that an arrival occurs, there is an asymptotic probability ν_{n-1} of observing n users in the system. On the other hand, out of state $n \leq m - 1$ (which has asymptotic probability π_n) the exponential servers provide a constant rate $n \cdot \mu$ to switch to state $n - 1$. Now the rate conservation law states that asymptotically

$$
\nu_{n-1} \cdot \frac{1}{\mathbb{E}(A)} = \pi_n \cdot n\mu
$$

which means that the probability flow from state $n - 1$ to state n equals the flow from n to $n - 1$. This yields

$$
\pi_n = \frac{\nu_{n-1}}{n\mu \cdot \mathbb{E}(A)} \tag{8.25}
$$

for $n = 1, \ldots, m - 1$. Finally we obtain π_0 by normalization, i.e.

$$\pi_0 = 1 - \sum_{n=1}^{\infty} \pi_n = 1 - \rho \cdot c - m\rho \sum_{n=1}^{m-1} \frac{\nu_{n-1}}{n}$$

This and formulae (8.25) and (8.24) collect all asymptotic probabilities.

Notes

The idea to analyze the M/G/1 queue via its embedded Markov chain has been presented in Kendall [49]. Earlier text book presentations for the M/G/1 and the GI/M/1 queue can be found in Cohen [28] or Çinlar [25]. The former contains further many special queues which are analyzed via embedded Markov chains as well. The GI/M/m queue has been examined in Kleinrock [50] and Asmussen [5]. The latter contains further an exact presentation of the rate conservation law.

For more examples of semi–Markovian queues see Cohen [28]. An application of the semi–Markov method to tandem queues is given in Breuer et al. [23].

Exercise 8.1 Assume exponential service times and a Poisson arrival process for the M/G/1 and GI/M/1 queue, respectively, and show that the results for the asymptotic distribution coincide with the results obtained for the M/M/1 queue.

Exercise 8.2 Verify equalities (8.8) and (8.9).

Exercise 8.3 Compute mean and variance of the asymptotic number of users in the system for the GI/M/1 queue. These may be expressed in terms of ξ. Derive further the mean sojourn time in the system.

Exercise 8.4 Verify equality (8.17).

Exercise 8.5 For an M/G/1 queue, show that the z–transform of the number of users which arrive during a service is given by

$$H(z) = \sum_{n=0}^{\infty} a_n z^n = B^*(\lambda - \lambda z)$$

for all $|z| < 1$, where B^* denotes the LST of the service time distribution. Show that the mean and the variance of this number are ρ and $\lambda^2 \mu_2(B)$, respectively, where $\mu_2(B)$ denotes the second moment of the service time distribution.

Exercise 8.6 Consider an M/G/1 queue with batch arrivals. Instead of single arrivals as in the ordinary M/G/1 queue, at every arrival instant the arrival consists of a batch of $n \in \mathbb{N}$ independent users with probability g_n. Arrival instants are distributed as a Poisson process with intensity $\lambda > 0$.
a) Define the z–transform $G(z) := \sum_{n=1}^{\infty} g_n z^n$ of the batch size distribution. Show that the z–transform of the number of users which arrive in an interval of length t is $N^*(t, z) = e^{-\lambda t(1 - G(z))}$.
b) Let $B^*(s)$ denote the LST of the service time distribution. Show that the z–transform of the number of users which arrive during a service is given by $B^*(\lambda - \lambda G(z))$.

Exercise 8.7 Let Q_n^d denote the number of users in the system after the nth departure and K_n the number of arrivals between the nth and the $n + 1$st departure. Justify the relation

$$D_{n+1} = (D_n - 1)^+ + K_n$$

where $a^+ = \max(0, a)$. As n tends to infinity, we obtain for $D := \lim_{n \to \infty} D_n$ and $K := \lim_{n \to \infty} K_n$ the equality

$$D = D - 1_{D>0} + K$$

in distribution. Take the expectation of D^2 in order to derive

$$\mathbb{E}(D) = \rho + \frac{\rho^2 \mu_2(B)}{2(1 - \rho)\mu_1^2(B)} \tag{8.26}$$

where $\mu_1^2(B)$ and $\mu_2(B)$ denote the squared first and the second moment of the service time distribution. Use the results from exercise 8.5. Why is this also the mean asymptotic number of users in the system? Equation (8.26) is known as the **Pollaczek–Khinchin mean value formula**.

Exercise 8.8 Show that the asymptotic mean number of users in an M/G/1 queue is minimal for deterministic service time distributions.

Exercise 8.9 Derive expression (8.20).

PART III

MATRIX–ANALYTIC METHODS

Chapter 9

PHASE–TYPE DISTRIBUTIONS

The memoryless property of the exponential distribution has been substantial for arriving at embedded Markov chains in chapter 8 when analyzing GI/M/1 and M/G/1 queues. The stationary distributions of these chains served as a foothold for a semi–regenerative analysis.

Our goal pursued in the next two chapters is to find generalizations beyond the exponential distribution and/or the Poisson process that are more versatile in their modelling capacity, but still allowing analyses of the respective queues by means of embedded Markov chains.

1. Motivation

In the present chapter, an extremely versatile class of distributions, the so–called phase–type or PH distributions, will be introduced. It is possible to approximate any distribution on the non–negative real numbers by a PH distribution, and the resulting queueing models can be analyzed almost as if we have dealt with the exponential distribution.

As a motivation, we begin with a practical example. Consider the **M/M/c/c+K queue**, which is defined as follows. Arrivals are modelled by a Poisson process with rate $\lambda > 0$. Service times are exponentially distributed with rate $\mu > 0$. There are c servers, and the capacity of the waiting room is K. That means that in total there is room for $c+K$ users in the system including the servers. If upon an arrival the system is filled, i.e. with $c + K$ users already in it, this arriving user is not admitted into the system. In this case we say that the arriving user is lost. Queueing systems with the possibility of such an event are thus called **loss system**s.

The queue described above is a simple Markov process with generator

$$
Q = \begin{pmatrix}
-\lambda & \lambda \\
\mu & -\lambda - \mu & \lambda \\
& 2\mu & -\lambda - 2\mu & \lambda \\
& & \ddots & \ddots & \ddots \\
& & & c\mu & -\lambda - c\mu & \lambda \\
& & & & \ddots & \ddots & \ddots \\
& & & & & c\mu & -\lambda - c\mu & \lambda \\
& & & & & & c\mu & -c\mu
\end{pmatrix}
$$

up to the first loss (all non–specified entries equal zero).

From a system administrator's point of view, the loss of a user is regarded as a bad event, and thus the question arises naturally how the distribution of the time up to the first loss might be expressed. However, the above description of the queueing process simply ignores loss events, as can be seen from the missing λ entries in the last line of the generator.

In order to include a possible loss event into our model of the queue, we add a new element to the state space and enlarge the generator as follows:

$$
Q' = \begin{pmatrix}
-\lambda & \lambda \\
\mu & -\lambda - \mu & \lambda \\
& \ddots & \ddots & \ddots \\
& & c\mu & -\lambda - c\mu & \lambda \\
& & & \ddots & \ddots & \ddots \\
& & & & c\mu & -\lambda - c\mu & \lambda \\
& & & & & c\mu & -\lambda - c\mu & \lambda \\
0 & & \cdots & & \cdots & & & 0
\end{pmatrix}
$$

again with all non–specified entries being zero. The first $m = c + K + 1$ states describe the number of users in the system, just as in the former generator Q. But now there is the possibility to enter another state $m + 1$ with rate λ from state m, obviously meaning exactly that a loss has occured. Since we want to observe the system only until the first loss, we choose the loss state $m + 1$ as an absorbing one. Thus all entries in the last line are zero.

Now the system administrator's question can be formulated mathematically as the distribution of the time until the Markov process with generator Q' enters the absorbing state $m + 1$. Exactly this problem is addressed (in a general form) by the concept of a phase–type distribution.

2. Definition and Examples

Definition 9.1 Let $\mathcal{X} = (X_t : t \geq 0)$ denote an homogeneous Markov process with finite state space $\{1, \ldots, m+1\}$ and generator

$$Q = \begin{pmatrix} T & \eta \\ 0 & 0 \end{pmatrix}$$

where T is a square matrix of dimension m, η a column vector and 0 the zero row vector of the same dimension. The **initial distribution** of $\mathcal{X}$ shall be the row vector $\tilde{\alpha} = (\alpha, \alpha_{m+1})$, with α being a row vector of dimension m. The first states $\{1, \ldots, m\}$ shall be transient, while the state $m+1$ is absorbing. Let $Z := \inf\{t \geq 0 : X_t = m+1\}$ be the random variable of the **time until absorption** in state $m+1$.

The distribution of Z is called **phase–type distribution** (or shortly **PH distribution**) with parameters (α, T). We write $Z \sim PH(\alpha, T)$. The dimension m of T is called the **order** of the distribution $PH(\alpha, T)$. The states $\{1, \ldots, m\}$ are also called **phases**, which gives rise to the name phase–type distribution.

Let 1 denote the column vector of dimension m with all entries equal to one. The first observations to be derived from the above definition are

$$\eta = -T1 \quad \text{and} \quad \alpha_{m+1} = 1 - \alpha 1$$

These follow immediately from the properties that the row sums of a generator are zero and the sum of a probability vector is one. The vector η is called the **exit vector** of the PH distribution. Now the distribution function and the density of a PH distribution are derived in

Theorem 9.2 Let $Z \sim PH(\alpha, T)$. Then the distribution function of Z is given by

$$F(t) := \mathbb{P}(Z \leq t) = 1 - \alpha e^{T \cdot t} 1 \tag{9.1}$$

for all $t \geq 0$, and the density function is

$$f(t) = \alpha e^{T \cdot t} \eta \tag{9.2}$$

for all $t > 0$. Here, the function $e^{T \cdot t} := \exp(T \cdot t) := \sum_{n=0}^{\infty} \frac{t^n}{n!} T^n$ denotes a **matrix exponential function.**

Proof: For the Markov process $\mathcal{X}$ with generator Q as given in definition 9.1 the equation

$$P(t) = \exp(Q \cdot t) = \begin{pmatrix} e^{T \cdot t} & 1 - e^{T \cdot t} 1 \\ 0 & 1 \end{pmatrix}$$

holds for the transition matrix $P(t)$ at every time $t \geq 0$. This implies

$$F(t) = \tilde{\alpha}e^{Q \cdot t}e_{m+1} = \alpha_{m+1} + \alpha \cdot (1 - e^{T \cdot t}1) = \alpha_{m+1} + \alpha 1 - \alpha e^{T \cdot t}1$$
$$= 1 - \alpha e^{T \cdot t}1$$

with e_{m+1} denoting the $m+1$st canonical base vector. For the density function we obtain

$$f(t) = F'(t) = -\alpha\frac{d}{dt}e^{T \cdot t}1 = -\alpha T e^{T \cdot t}1 = \alpha e^{T \cdot t}(-T1) = \alpha e^{T \cdot t}\eta$$

which was to be proven.

$\square$

A first consequence is $F(0) = \alpha_{m+1}$, which is also clear from definition 9.1. An important question to be examined is when a phase–type distribution is **non–defective**, i.e. what the conditions for $F(\infty) = \lim_{t \to \infty} F(t) = 1$ are. This is answered in

Theorem 9.3 Let F denote a $PH(\alpha, T)$ distribution function. F is non–defective, i.e. $F(\infty) = 1$ for all α, if and only if T is invertible. In this case, $(-T^{-1})_{ij}$ is the expected total time spent in state j given that the process $\mathcal{X}$ started in state i.

Proof: Let E_{ij} denote the expected total time spent in state j given that the process $\mathcal{X}$ started in i. Define $E = (E_{ij})_{i,j \leq m}$ as the respective matrix of expectations.

First we assume that $F(\infty) = 1$ for all α, i.e. that F is non–defective. This means that with probability one there is an absorption from any initial state i. This implies for E to be finite, i.e.

$$E_{ij} < \infty \qquad \text{for all } i, j \in \{1, \dots, m\}$$

Conditioning on the first state visited after i yields the relations

$$E_{ij} = \sum_{k \neq i} \frac{T_{ik}}{-T_{ii}}E_{kj} \quad \text{for all } i \neq j$$

$$E_{ii} = \frac{1}{-T_{ii}} + \sum_{k \neq i} \frac{T_{ik}}{-T_{ii}}E_{ki}$$

In matrix notation, this is expressed as $TE = -I$, with I denoting the identity matrix. Hence we obtain $E = -T^{-1}$, which was to be proven.

Now we assume that T is invertible. Define the vector $\Phi(x) = \exp(Tx)\mathbf{1}$ for all $x \geq 0$. The numbers $\Phi_i(x)$ are the probabilities that the process $\mathcal{X}$ is in one of the states $\{1, \ldots, m\}$ after time x given that the initial state was i. Hence

$$\Phi_i(x) \in [0, 1] \qquad \text{for all } i \in \{1, \ldots, m\}$$

Further the equation

$$e^{T \cdot t} = I + \int_0^t T e^{T \cdot x} \, dx$$

holds as can be seen by differentiating both sides and acknowledging $e^{T \cdot 0} = I$. Multiplying by T^{-1} from the left side and by $\mathbf{1}$ from the right side yields

$$T^{-1} \Phi(t) = T^{-1} e^{T \cdot t} \mathbf{1} = T^{-1} \mathbf{1} + \int_0^t e^{T \cdot x} \, dx \mathbf{1}$$

As all entries of $\Phi(t)$ are finite, we obtain for t tending to infinity

$$\lim_{t \to \infty} \int_0^t e^{T \cdot x} \, dx < \infty$$

in an entry–wise meaning. But the values $(e^{T \cdot x})_{ij}$ are simply the probability that the process $\mathcal{X}$ is in state j at time x given that it started in state i. Hence we obtain

$$E_{ij} = \int_0^\infty (e^{T \cdot x})_{ij} \, dx < \infty \qquad \text{for all } i, j \in \{1, \ldots, m\}$$

which means that all states $j \in \{1, \ldots, m\}$ are transient. Thus an absorption in state $m + 1$ is certain regardless of the initial distribution, which was to be proven.

□

From now on T shall be assumed to be invertible. In order to show the versatility of the phase–type concept, we shall give a few examples below. Important characteristics for distributions are their moments. Given a distribution function F, its nth **moment** (if existing) is given by

$$M_n(F) := \int_0^\infty t^n dF(t)$$

Clearly, the first moment is the mean of the distribution. The nth moment of the exponential distribution with parameter λ is given by $M_n = n!/(\lambda^n)$ (see exercises). Another important characteristic is the so–called squared coefficient of variation, defined by

$$C_V(F) := \mathbb{V}ar(F)/\mathbb{E}(F)^2$$

with $\mathbb{V}ar(F)$ denoting the variance of F. For any exponential distribution this equals one. The values of the squared coefficient of variation will explain the names for the hypo- and hyper–exponential distributions introduced below.

Example 9.4 Erlang distribution

A well–known distribution within the family of Gamma distributions is the so–called **Erlang distribution**. An Erlang distribution E_n^λ with n degrees of freedom (or stages) and parameter λ is the distribution of the sum of n exponential random variables with parameter λ. It has the density function

$$f(t) = \frac{\lambda^n}{(n-1)!} t^{n-1} e^{-\lambda t}$$

for all $t \geq 0$. Its interpretation as a succession of n exponential distributions with rate λ each can be illustrated graphically as in

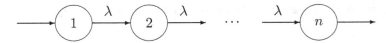

Figure 9.1.   Erlang distribution

Here we see that an Erlang distribution can be represented as the holding time in the transient state set $\{1, \ldots, n\}$ of a Markov chain with absorbing state $n+1$ where the only possible transitions occur from a state k to the next state $k+1$ (for $k = 1, \ldots, n$), with rate λ each. In terms of our definition 9.1, we have a PH representation

$$\alpha = (1, 0, \ldots, 0), \quad T = \begin{pmatrix} -\lambda & \lambda & & \\ & \ddots & \ddots & \\ & & -\lambda & \lambda \\ & & & -\lambda \end{pmatrix} \quad \text{and} \quad \eta = \begin{pmatrix} 0 \\ \vdots \\ 0 \\ \lambda \end{pmatrix}$$

with all non–specified entries in T being zero.

The mean of an Erlang distribution with n degrees of freedom and parameter λ is n/λ, while its squared coefficient of variation is $1/n$, i.e. less than one if $n > 1$ (see exercises). This explains the name **hypo–exponential distribution** appearing in the next example.

Example 9.5 Generalized Erlang distribution

A slight generalization of the Erlang distribution is obtained if one admits the exponential stages to have different parameters. Then we talk about a generalized Erlang (or a hypo–exponential) distribution. The representation as a PH distribution results in the figure

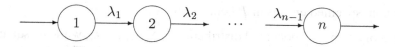

Figure 9.2. Generalized Erlang distribution

and leads to a PH representation

$$\alpha = (1, 0, \ldots, 0), \quad T = \begin{pmatrix} -\lambda_1 & \lambda_1 & & \\ & \ddots & \ddots & \\ & & -\lambda_{n-1} & \lambda_{n-1} \\ & & & -\lambda_n \end{pmatrix} \quad \text{and} \quad \eta = \begin{pmatrix} 0 \\ \vdots \\ 0 \\ \lambda_n \end{pmatrix}$$

with all non–specified entries in T being zero. For this family of distributions, a closed formula for the density function is already rather complex.

Example 9.6 Hyper–exponential distribution
A **hyper–exponential distribution** is a finite mixture of $n \in \mathbb{N}$ exponential distributions with different parameters λ_k $(k = 1, \ldots, n)$. Its density function is given as

$$f(t) = \sum_{k=1}^{n} q_k \lambda_k e^{-\lambda_k t}$$

with proportions $q_k > 0$ satisfying $\sum_{k=1}^{n} q_k = 1$. A graphical representation of this distribution is

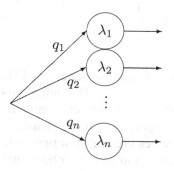

Figure 9.3. Hyper–exponential distribution

This leads to a PH representation by

$$\alpha = (\pi_1, \ldots, \pi_n), \quad T = \begin{pmatrix} -\lambda_1 & & \\ & \ddots & \\ & & -\lambda_n \end{pmatrix} \quad \text{and} \quad \eta = \begin{pmatrix} \lambda_1 \\ \vdots \\ \lambda_n \end{pmatrix}$$

with all non–specified entries in T being zero.

The mean of a hyper–exponential distribution is $\sum_{i=1}^{n} \pi_i/\lambda_i$, while its squared coefficient of variation is always larger than one if $n > 1$. This explains the name hyper–exponential distribution.

Example 9.7 Cox distribution
A more complex example of the classical families of distributions are the **Cox distribution**s. These are generalized Erlang distributions with preemptive exit options. A Coxian random variable measures the holding time within the box depicted as

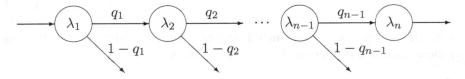

Figure 9.4. Cox distribution

A Cox distribution can be described as a special PH distribution with parameters $\alpha = (1, 0, \ldots, 0)$ and

$$
T = \begin{pmatrix} -\lambda_1 & q_1\lambda_1 & & \\ & \ddots & \ddots & \\ & & -\lambda_{n-1} & q_{n-1}\lambda_{n-1} \\ & & & -\lambda_n \end{pmatrix}, \qquad \eta = \begin{pmatrix} (1-q_1)\lambda_1 \\ \vdots \\ (1-q_{n-1})\lambda_{n-1} \\ \lambda_n \end{pmatrix}
$$

for which all non–specified entries in T are zero.

As we have seen in example 9.6, the set of transient states in a PH distribution may fall apart into several communication classes. The definition of a phase–type distribution leaves open the possibility of a transient communication class which cannot be entered because the respective initial probabilities contained in the row vector α are zero. Such states are called **superfluous**, since the Markov process $\mathcal{X}$ defining the PH distribution will never be in such a state. As we will obtain the same distribution of the time until absorption if we leave out superfluous states, we shall from now on (unless stated otherwise) assume that there are no superfluous states in the definition of a phase–type distribution.

3. Moments

A good means to determine the moments of a distribution is the **Laplace–Stieltjes transform** (or shortly **LST**, see appendix). This is derived in

Theorem 9.8 *The LST of a phase–type distribution $F = PH(\alpha, T)$ is given by*

$$\Phi(s) := \int_0^\infty e^{-st} \, dF(t) = \alpha_{m+1} + \alpha(s \cdot I - T)^{-1}\eta$$

for all $s \in \mathbb{C}$ with $Re(s) \geq 0$.

Proof: For $s = 0$ the statement is obvious. Be $s \neq 0$ and $Re(s) \geq 0$. Integration by parts yields

$$\int_0^\infty e^{-st} e^{T \cdot t} \, dt = \frac{-1}{s} \left(\left[e^{-st} e^{T \cdot t}\right]_0^\infty - \int_0^\infty e^{-st} e^{T \cdot t} T \, dt \right)$$

As T is assumed to be invertible, we know from the proof of theorem 9.3 that $\int_0^\infty e^{T \cdot t} \, dt < \infty$ entry–wise, and hence $\lim_{t \to \infty} e^{T \cdot t} = 0$, also entry–wise. This yields

$$\int_0^\infty e^{-st} e^{T \cdot t} \, dt = \frac{1}{s} \cdot I + \frac{1}{s} \cdot \int_0^\infty e^{-st} e^{T \cdot t} \, dt \, T$$

which implies

$$\int_0^\infty e^{-st} e^{T \cdot t} \, dt \, (s \cdot I - T) = I$$

and hence

$$\int_0^\infty e^{-st} e^{T \cdot t} \, dt = (s \cdot I - T)^{-1} \qquad \text{for all } s \neq 0 \text{ with } Re(s) \geq 0.$$

This leads to

$$\Phi(s) = \int_0^\infty e^{-st} \, dF(t) = \alpha_{m+1} + \int_0^\infty e^{-st} \alpha e^{T \cdot t} \eta \, dt$$

$$= \alpha_{m+1} + \alpha \int_0^\infty e^{-st} e^{T \cdot t} \, dt \, \eta = \alpha_{m+1} + \alpha(s \cdot I - T)^{-1}\eta$$

which is the statement.
□

Corollary 9.9 *Let $Z \sim PH(\alpha, T)$. The moments of Z are given by*

$$\mathbb{E}(Z^n) = (-1)^n \cdot n! \cdot \alpha T^{-n} \mathbf{1}$$

for all $n \in \mathbb{N}$.

Proof: Because of

$$\frac{d^n}{ds^n}(s \cdot I - T)^{-1} = (-1)^n \cdot n! \cdot (s \cdot I - T)^{-(n+1)}$$

we obtain

$$\mathbb{E}(Z^n) = (-1)^n \cdot \frac{d^n}{ds^n} \Phi(s)|_{s=0}$$
$$= n! \cdot \alpha(s \cdot I - T)^{-(n+1)}|_{s=0} \eta = n! \cdot \alpha(-T)^{-n}(-T)^{-1}(-T)\mathbf{1}$$
$$= n! \cdot \alpha(-T)^{-n}\mathbf{1}$$

$\square$

In the next chapter it will be shown in corollary 10.12 that another expression for the mean is given by $\mathbb{E}(Z) = (\pi\eta)^{-1}$ with $\pi(T + \eta\alpha) = 0$.

4. Closure Properties

A useful advantage of phase–type distributions is the fact that certain compositions of PH distributions result in PH distributions again. This means that the class of PH distributions is closed under these compositions. For PH distributed random variables Z_1 and Z_2 we will show **closure properties** for the compositions $Z_1 + Z_2$ (convolution), $pZ_1 + (1-p)Z_2$ with $p \in [0, 1]$ (mixture), and $\min(Z_1, Z_2)$.

Theorem 9.10 *Let $Z_i \sim PH(\alpha^{(i)}, T^{(i)})$ of order m_i for $i = 1, 2$. Then $Z = Z_1 + Z_2 \sim PH(\alpha, T)$ of order $m = m_1 + m_2$ with representation*

$$\alpha_k = \begin{cases} \alpha_k^{(1)}, & 1 \leq k \leq m \\ \alpha_{m_1+1}^{(1)} \cdot \alpha_{k-m_1}^{(2)}, & m_1 + 1 \leq k \leq m \end{cases}$$

and

$$T = \begin{pmatrix} T^{(1)} & \eta^{(1)}\alpha^{(2)} \\ \mathbf{0} & T^{(2)} \end{pmatrix}$$

where $\eta^{(1)} = -T^{(1)}\mathbf{1}_{m_1}$ and $\mathbf{0}$ denotes a zero matrix of appropriate dimension.

Proof: By definition, Z_i is the random variable of the time until absorption in a Markov process $\mathcal{X}_i$ with transient states $\{1, \ldots, m_i\}$ and an absorbing state which shall be denoted by e_i in this proof. The transition rates of $\mathcal{X}_i$ within the set of transient states are given by the matrix $T^{(i)}$ and the absorption rates from the transient states to the absorbing state are given by the vector $\eta^{(i)}$.

Then the random variable $Z = Z_1 + Z_2$ is the total time duration of first entering e_1 and then e_2 in the Markov process which is structured as follows:

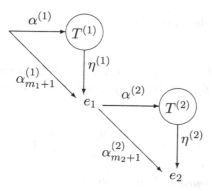

Figure 9.5. Convolution of two PH distributions

Here the point e_1 is not a state of the Markov process described above but only an auxiliary construction aid for a better illustration. In particular there is no holding time in e_1. The only absorbing state in the Markov process constructed above is e_2.

With probability $\alpha^{(1)}_{m_1+1}$ we enter the first absorbing state e_1 immediately, while the vector $\alpha^{(1)}$ contains the probabilities that we first enter the set of transient states of X_1. In the latter case, the matrix $T^{(1)}$ and then the vector $\eta^{(1)}$ determine the time until the first absorption in e_1.

After having reached e_1, the chain immediately (i.e. with no holding time in e_1) proceeds to the second stage, which is completely analogous to the first. With probability $\alpha^{(2)}_{m_2+1}$ we enter the second absorbing state e_2 immediately, while the vector $\alpha^{(2)}$ contains the probabilities that we first enter the set of transient states of X_2. In the latter case, the matrix $T^{(2)}$ and then the vector $\eta^{(2)}$ determine the time until the first absorption in e_2.

Thus we get to the second absorbing state e_2 immediately with probability $\alpha^{(1)}_{m_1+1} \cdot \alpha^{(2)}_{m_2+1}$. There are transient states $\{1, \ldots, m_1, m_1+1, \ldots, m_1+m_2\}$. The first m_1 of these are reached with probabilities $\alpha_1, \ldots, \alpha_{m_1}$, while the last m_2 of these states can only be reached via an immediate first absorption in e_1 and thus with probabilities $\alpha^{(1)}_{m_1+1} \cdot \alpha^{(2)}_i$ for $i = 1, \ldots, m_2$. This explains the expression for α.

In order to explain the structure of T, we observe first that there is no path from the second set of transient states to the first, whence the lower left entry of T is zero. The diagonal entries of T describe the transition rates within the two sets of transient states, respectively, and thus are given by $T^{(1)}$ and $T^{(2)}$. The only way to get from the first to the second set of transient states is the path via e_1 for which we first need the rates given in $\eta^{(1)}$ and then the probabilities contained in $\alpha^{(2)}$. Hence the upper right entry of T.

$\square$

Theorem 9.11 *Let $Z_i \sim PH(\alpha^{(i)}, T^{(i)})$ of order m_i for $i = 1, 2$, as well as $p \in [0, 1]$. Then $Z = pZ_1 + (1 - p)Z_2 \sim PH(\alpha, T)$ of order $m = m_1 + m_2$ with representation*

$$\alpha = (p \cdot \alpha^{(1)}, (1 - p) \cdot \alpha^{(2)}) \quad and \quad T = \begin{pmatrix} T^{(1)} & \mathbf{0} \\ \mathbf{0} & T^{(2)} \end{pmatrix}$$

where $\mathbf{0}$ denote zero matrices of appropriate dimensions.

Proof: Going along the line of reasoning of the last proof, we observe that Z is equal to Z_1 with probability p and equal to Z_2 with probability $1 - p$. Hence we obtain the following construction of a Markov process:

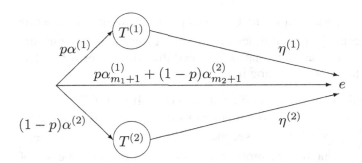

Figure 9.6. Mixture of two PH distributions

Here, we enter the first set of transient states with probabilities $p \cdot \alpha_i^{(1)}$ for $i = 1, \ldots, m_1$ and the second set with probabilities $(1 - p) \cdot \alpha_i^{(2)}$ for phases $i = m_1 + 1, \ldots, m_2$. This explains the expression for α.

From either of these sets we proceed with transition matrices $T^{(i)}$ and exit vectors $\eta^{(i)}$, $i = 1, 2$, in order to reach the absorbing state e. There is no path from one set of transient states to the other, which explains the structure of T.

The absorbing state e can be reached immediately (i.e without entering any transient state) with probability $p\alpha_{m_1+1}^{(1)} + (1 - p)\alpha_{m_2+1}^{(2)}$.

$\square$

In order to formulate the next theorem, we first need to define the so–called Kronecker compositions of matrices. Let $A = (a_{ij})$ and $B = (b_{ij})$ denote

$n_1 \times m_1$ and $n_2 \times m_2$ matrices, respectively. The **Kronecker product** of A and B is defined as the $(n_1 \cdot n_2) \times (m_1 \cdot m_2)$ matrix $A \otimes B$ with entries

$$(A \otimes B)_{(i_1,i_2),(j_1,j_2)} := a_{i_1,j_1} \cdot b_{i_2,j_2}$$

for all $1 \leq i_k \leq n_k$ and $1 \leq j_k \leq m_k$, $k = 1, 2$. As a block matrix we can write

$$A \otimes B = \begin{pmatrix} a_{11}B & \cdots & a_{1m_1}B \\ \vdots & & \vdots \\ a_{n_11}B & \cdots & a_{n_1m_1}B \end{pmatrix}$$

If A and B are square matrices, i.e. $n_k = m_k$ for $k = 1, 2$, then the **Kronecker sum** $A \oplus B$ of the matrices A and B is defined as

$$A \oplus B := A \otimes I_2 + I_1 \otimes B$$

with I_k denoting the $n_k \times n_k$ identity matrix for $k = 1, 2$.

Example 9.12 Let $n_1 = m_1 = 1$ and $n_2 = m_2 = 2$. If $A = -\lambda$ and

$$B = \begin{pmatrix} -\mu & \mu \\ 0 & -\mu \end{pmatrix}, \quad \text{then} \quad A \oplus B = \begin{pmatrix} -(\lambda + \mu) & \mu \\ 0 & -(\lambda + \mu) \end{pmatrix}$$

is an explicit expression for the Kronecker sum of A and B.

Theorem 9.13 *Let $Z_i \sim PH(\alpha^{(i)}, T^{(i)})$ of order m_i for $i = 1, 2$ and define $Z = \min(Z_1, Z_2)$. Then $Z \sim PH(\alpha, T)$ of order $m = m_1 \cdot m_2$ with representation*

$$\alpha = \alpha^{(1)} \otimes \alpha^{(2)} \quad \text{and} \quad T = T^{(1)} \oplus T^{(2)}$$

in terms of the Kronecker compositions.

Proof: For $i = 1, 2$, the random variables Z_i are the times until absorption in the Markov processes $\mathcal{X}_i = (X_t^{(i)} : t \geq 0)$ where the initial distributions for the transient states are $\alpha^{(i)}$ and the transition rates among the transient states are given by $T^{(i)}$. Thus we can determine Z if we start running $\mathcal{X}_1$ and $\mathcal{X}_2$ concurrently and stop whenever the first of the two processes enters the absorbing state. We will show that the two–dimensional Markov process $\mathcal{X}$ depicted as in the figure below has the same time until absorption as the first absorption of the concurrent processes $\mathcal{X}_1$ and $\mathcal{X}_2$.

The state space of $\mathcal{X}$ shall be

$$E = \{(i, j) : 1 \leq i \leq m_1, 1 \leq j \leq m_2\} \cup \{e\}$$

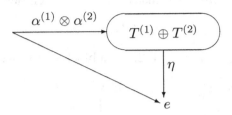

Figure 9.7.   Superposition of two PH distributions

where e is the absorbing state and all other states are transient. We will keep in mind the interpretation that $X_t = (i, j)$ means $X_t^{(1)} = i$ and $X_t^{(2)} = j$ for all transient states i, j. The exit vector η of dimension $m_1 \cdot m_2$ has entries $\eta_{ij} = \eta_i^{(1)} + \eta_j^{(2)}$ for all $i \leq m_1$ and $j \leq m_2$.

Since we start the processes $\mathcal{X}_1$ and $\mathcal{X}_2$ independently, we clearly have an initial distribution

$$\mathbb{P}(X_0 = (i, j)) = \mathbb{P}(X_0^{(1)} = i) \cdot \mathbb{P}(X_0^{(2)} = j) = \alpha_i^{(1)} \cdot \alpha_j^{(2)}$$

which explains the expression for α. If the process $\mathcal{X}$ is in state (i, j), this means that the exponential holding times in the states i (for $\mathcal{X}_1$) and j (for $\mathcal{X}_2$) are running concurrently. According to lemma 3.2 they almost certainly do not stop at the same time instant. Thus $\mathcal{X}$ has only state transitions from (i, j) to either (i, h) or (k, j). These occur with transition rates $T_{jh}^{(2)}$ or $T_{ik}^{(1)}$, respectively, if h and k are transient states. According to lemma 3.2 the holding time in state (i, j) is exponential with parameter $-(T_{ii}^{(1)} + T_{jj}^{(2)})$. This explains the structure of T. The values of η are readily verified by

$$\eta_{ij} = -(T\mathbf{1}_{m_1 m_2})_{(i,j)} = -\left(T_{ii}^{(1)} + T_{jj}^{(2)} + \sum_{h \neq j} T_{jh}^{(2)} + \sum_{k \neq i} T_{ik}^{(1)} \right)$$

$$= -\sum_{k=1}^{m_1} T_{ik}^{(1)} - \sum_{h=1}^{m_2} T_{jh}^{(2)} = \eta_i^{(1)} + \eta_j^{(2)}$$

Thus we can see that $Z = \min(Z_1, Z_2)$ is the time until absorption in the Markov process $\mathcal{X}$.

□

Example 9.4 and theorem 9.11 already suffice to prove the following powerful

Theorem 9.14 *The class of phase–type distributions is dense (in terms of weak convergence) within the class of all distributions on $\mathbb{R}_0^+$.*

Proof: Let $F : \mathbb{R}_0^+ \to [0,1]$ denote any non–negative distribution function. Since it is bounded, monotone and right–continuous, we can approximate F by a step function G with countably many jumps at $(t_n : n \in \mathbb{N}_0)$, where $t_n < t_{n+1}$ for all $n \in \mathbb{N}$. The error $\varepsilon > 0$ of approximation can be chosen arbitrarily small such that $|F(t) - G(t)| < \varepsilon$ holds for all $t \geq 0$.

If $t_0 = 0$, i.e. if there is a jump of G at zero, we can write

$$G = p_0 \cdot \delta_0 + (1 - p_0) \cdot \tilde{G}$$

with $p_0 = G(0)$ and $\tilde{G} = (G-G(0))/(1-p_0)$. The Dirac distribution function δ_0 is a phase–type distribution with $m = 0$ and $\alpha_{m+1} = 1$. In view of example 9.4 and theorem 9.11, it now suffices to show that we can approximate the function $\tilde{G}$ by a finite mixture of Erlang distributions. First we find a truncation point T of $\tilde{G}$ such that $G(T) > 1 - \varepsilon$. Then there is a number $N \in \mathbb{N}$ such that $t_n > T$ for all $n > N$. Thus $\tilde{G}$ can be approximated by

$$H = \sum_{n=1}^{N-1} (\tilde{G}(t_n) - \tilde{G}(t_{n-1})) \cdot \delta_{t_n} + (1 - \tilde{G}(t_N)) \cdot \delta_{t_N}$$

with an error bounded by ε.

For every $n = 1, \ldots, N$ we approximate the Dirac distribution δ_{t_n} by a suitable Erlang distribution. This possible because of the following argument: The variance of an Erlang distribution $E_k^{k\lambda}$ of order k with parameter $k \cdot \lambda$ is given by $(k \cdot \lambda^2)^{-1}$ (see exercises) and thus tends to zero as k grows larger. Since the mean of such an Erlang distribution is $1/\lambda$ (see exercises), Chebyshev's inequality tells us that the sequence $(E_k^{k\lambda} : k \in \mathbb{N})$ converges in probability (and hence weakly) towards δ_{t_n} if we chose $\lambda = 1/t_n$. This means that there is a number $K \in \mathbb{N}$ such that the distribution function H_n of an E_K^{K/t_n} distribution satisfies $|H_n(t) - \delta_{t_n}(t)| < \varepsilon$ for all $t \geq 0$.

If we pursue the above approximation method for every $n = 1, \ldots, N$ and define

$$\tilde{H} = \sum_{n=1}^{N-1} (\tilde{G}(t_n) - \tilde{G}(t_{n-1})) \cdot H_n + (1 - \tilde{G}(t_N)) \cdot H_N$$

then we obtain an approximation bound $|H - \tilde{H}| < \varepsilon$. According to example 9.4 and theorem 9.11 the distribution $\tilde{H}$ is phase–type.

In summary, we have approximated F by $p_0 \cdot \delta_0 + (1 - p_0) \cdot \tilde{H}$ with an approximation bound of $3 \cdot \varepsilon$. This proves our statement.

$\square$

Notes

Phase–type distributions have been introduced in Neuts [62] as a generalization of the Erlang and hyper–exponential distribution. A classical introduction to phase–type distributions is given in Neuts [65]. Statistical methods for fitting PH distributions are given in Asmussen et al. [7]. Phase–type distributions with infinitely many phases are introduced in Shi et al. [78], and Shi and Liu [79].

The name of a superfluous state as well as the motivating example at the beginning of the section have been taken from Latouche and Ramaswami [52].

The proofs of the closure properties have been chosen to be as constructive and illustrating as possible. More classical proofs via comparison of the Laplace transforms can be found in Neuts [65].

Exercise 9.1 Show that the density $f(t)$ of a phase–type distribution function is strictly positive for all $t > 0$.

Exercise 9.2 Compute the Laplace–Stieltjes transform, all moments as well as the squared coefficient of variation for the exponential, the Erlang, and the hyper–exponential distribution.

Exercise 9.3 Use the results from exercise 9.2 in order to show that the Erlang distribution with parameter $n\lambda$ and n degrees of freedom is the n–fold convolution of an exponential distribution with parameter $n\lambda$. Employ this result for a simple proof of formula (6.2).

Exercise 9.4 Consider two machines running independently at the same time. The one has a life time which is distributed as a generalized Erlang with two stages and parameters λ_1 and λ_2. The other machine's life time has a hyper–exponential distribution with density $f(t) = p \cdot \mu_1 e^{-\mu_1 t} + (1 - p) \cdot \mu_2 e^{-\mu_2 t}$. As soon as a machine fails, it is given to repair which takes an exponentially distributed amount of time with parameter κ. After repair it starts working immediately. Determine the distribution of the time until both machines are broken down.

Exercise 9.5 Consider the M/PH/k queue with Poisson input and phase–type service time distribution for all its k servers. Derive a description for this queue in terms of a $(k + 1)$–dimensional Markov process.

Exercise 9.6 Find the stationary distribution for the M/PH/1 queue.

Chapter 10

MARKOVIAN ARRIVAL PROCESSES

In this chapter we are going to generalize the concept of a Poisson process in three steps. We shall arrive thereby at the class of so–called Batch Markovian Arrival Processes (shortly BMAPs) that comprises a great variety of processes and so provides much more realistic modelling tools than the class of processes considered so far can offer. The big advantage of the very procedure of generalizing basic Markovian concepts is that it essentially keeps a lot of Markovian behaviour.

1. The PH renewal process

The Poisson process is a renewal process with exponentially distributed renewal intervals (see example 6.5). In the last section we have introduced a powerful generalization of the exponential distribution. As a first step to go beyond the Poisson process, it seems natural to replace the exponential distribution of the renewal intervals by a phase–type distribution. This leads us immediately to the class of PH renewal processes. To be precise, a renewal process $\mathcal{N} = (N_t : t \in \mathbb{R}_0^+)$ with phase–type distributed renewal intervals is called a **PH renewal process**.

While a concrete determination of the renewal function or the time–dependent behaviour of an arbitrary renewal process is hard to derive, we have seen in example 6.5 that in the case of exponential renewal intervals, i.e. for the Poisson process, simple expressions can be derived. Since an exponential distribution is a phase–type distribution of order $m = 1$, a PH renewal process clearly generalizes the Poisson process. We want to show that the behaviour of PH renewal processes can still be described by rather simple expressions.

To this aim, we first will give a Markovian description of a PH renewal process. For a phase–type distribution $PH(\alpha, T)$, the remaining time Z until absorption, given that the current phase is i, can be expressed by

$$\mathbb{P}(Z \leq t | X_0 = i) = 1 - e_i e^{T \cdot t} \mathbf{1}$$

and thus has the same form as a $PH(e_i, T)$ distribution, with e_i denoting the ith canonical row base vector. This is the **conditional memoryless property** of the phase–type distribution.

If we keep track of the current phase of the PH distribution during renewal intervals, then we obtain a Markovian description $(\mathcal{N}, \mathcal{J})$ of the PH renewal process $\mathcal{N}$. This is derived as follows (cf. notations in definition 1). Clearly, the state space of $\mathcal{Y} = (\mathcal{N}, \mathcal{J})$ is $E = \mathbb{N}_0 \times \{1, \ldots, m\}$, with N_t denoting the number of arrivals until time t and J_t being the current phase at time t.

The holding times depend only on the current phase of the PH distribution, i.e. $\lambda_{n,i} = \lambda_i$ for all $n \in \mathbb{N}_0$. Since $\mathcal{N}$ is a renewal process, the state transitions of the embedded Markov chain X are restricted by $p_{n,i;m,j} = 0$ for $m < n$ or $m > n + 1$.

Thus there are state transitions from (m, i) to $(m + 1, j)$ or to (m, j), which are called transitions with or without arrivals, respectively. For **transitions without arrivals**, there is no renewal event and hence no absorption for the PH distribution, which means that these transitions are described by the parameter matrix T of the PH distribution of the renewal intervals. For **transitions with arrivals**, we observe the following dynamics.

Being in phase i, there is an absorption (hence a renewal event) with rate η_i. After that, a new PH–distributed renewal interval begins and a phase is chosen according to the initial phase distribution α. Hence transitions with arrivals are determined by the rate matrix $A := \eta \alpha$ with $\eta := -T\mathbf{1}$.

After ordering the state space $E = \mathbb{N}_0 \times \{1, \ldots, m\}$ lexicographically, we can write the infinitesimal generator matrix G of the Markov process $\mathcal{Y} = (\mathcal{N}, \mathcal{J})$ as a block matrix

$$G = \begin{pmatrix} T & A & & \\ & T & A & \\ & & T & A \\ & & & \ddots & \ddots \end{pmatrix}$$

with the non–specified blocks being zero matrices, and $A := \eta \alpha$.

Before analyzing the behaviour of this process, we shall generalize its structure two steps further. Then we will pursue an analysis from a more general viewpoint.

2. From PH renewal processes to MAPs

An essential feature of the PH renewal process is that immediately after an arrival (i.e. a renewal event) the phase distribution always is α. This makes it a real renewal process with iid renewal intervals. However, in modern communication systems like the internet or other computer networks there may be strong correlations between subsequent inter–arrival times. Thus it is a natural idea to introduce a dependence between the subsequent renewal intervals. This can be done without changing the block structure of the generator.

Writing down the row vectors of the matrix A, we observe

$$A = \begin{pmatrix} \eta_1 \cdot \alpha \\ \vdots \\ \eta_m \cdot \alpha \end{pmatrix}$$

meaning that the row entries differ only by a scalar η_i and thus the new phase after an arrival is chosen independently of the phase immediately before that arrival.

If we relax this restriction, we arrive at a new matrix

$$A' = \begin{pmatrix} \eta_1 \cdot \alpha_1 \\ \vdots \\ \eta_m \cdot \alpha_m \end{pmatrix}$$

with the only requirement that $\alpha_i 1 = 1$ for all $i = 1, \ldots, m$. Here the phase distribution after an arrival depends on the phase immediately before that arrival. However, the requirement $\alpha_i 1 = 1$ in connection with the fact that $\eta = -T1$ simply restates the observation that the row entries of a generator matrix sum up to zero.

Thus there is no real restriction in choosing A'. If we denote $D_0 := T$ and $D_1 := A'$ as usually done in the literature, we arrive at a generator

$$G = \begin{pmatrix} D_0 & D_1 & & \\ & D_0 & D_1 & \\ & & D_0 & D_1 \\ & & & \ddots & \ddots \end{pmatrix}$$

with no restrictions on the matrices D_0 and D_1 except that G be a generator matrix and D_1 is non–negative. A Markov process with such a generator is called **Markovian Arrival Process** (or shortly **MAP**). These arrival processes play an important part in today's queueing models.

Differing from PH renewal processes as a special case, MAPs are not renewal but semi–Markov processes (see section 2). The methods we will employ for their analysis still allow a further generalization, which also is motivated by observations of traffic in modern communication networks.

3.   From MAPs to BMAPs

In a computer network it is not uncommon that a client sends various jobs to the server at the same time. After being sent they are treated as separate entities. For our modelling tools concerning the arrival process this means that we observe several arrivals at the same time. These are called **batch arrivals** or arrivals in batches.

It is a simple matter to include this into our concept of Markovian arrival processes. We observe that for MAPs the matrix D_0 on the main block diagonal contains the transition rates without arrivals. The matrix D_1 on the first upper block diagonal contains transition rates with single arrivals. The lexicographic order of the state space implies that any (positive) entry in the kth upper block diagonal of the generator matrix G would be a transition rate for an arrival of k jobs at the same time.

Hence a natural extension of MAPs which include the possibility of batch arrivals are Markov processes with a generator of the block structure

$$G = \begin{pmatrix} D_0 & D_1 & D_2 & D_3 & \cdots \\ & D_0 & D_1 & D_2 & \ddots \\ & & D_0 & D_1 & \ddots \\ & & & \ddots & \ddots \end{pmatrix}$$

again with the non–specified blocks (i.e. all entries in the lower block diagonals) being zero matrices. Such processes are called **Batch Markovian Arrival Processes** (or shortly **BMAPs**). A matrix D_k contains the transition rates for a batch arrival of size k, i.e. the event that k jobs arrive at the same time. The sequence $\Delta = (D_n : n \in \mathbb{N}_0)$ uniquely determines the generator G and thus the BMAP. We call it the **characterizing matrix sequence** or the sequence of **characterizing matrices** for the BMAP.

The entry $D_{k;ij}$ with $k \leq 1$ and $i, j \leq m$ indicates the transition rate for a batch arrival of size k occuring in connection with a phase transition from i to j. The entry $D_{0;ij}$ with $i \neq j \leq m$ indicates the transition rate for a phase transition from i to j without an arrival. Finally, the entry $D_{0;ii}$ with $i \leq m$ indicates the negative parameter of the exponential holding time in any state (n, i) with $n \geq 0$ and $i \leq m$.

Clearly, MAPs and thus PH renewal processes are special cases of BMAPs. The following examples further show the great versatility of the BMAP concept.

Example 10.1 Markov–modulated Poisson process
One of the most prominent examples of a MAP is the **Markov–modulated Poisson process** (or shortly: **MMPP**). It is a **doubly stochastic Poisson process** in the following sense: First we define an underlying (or governing) Markov process $\mathcal{X}$ with a finite number m of states. This is sometimes referred to as the **environment process**. Denote its generator by R. Depending on the state of $\mathcal{X}$, single arrivals occur with one of the rates $\lambda_1, \ldots, \lambda_m$. Define the diagonal matrix $\Lambda = diag(\lambda_1, \ldots, \lambda_m)$. Then we obtain a BMAP specification of the MMPP by setting $D_0 = R - \Lambda$, $D_1 = \Lambda$, and $D_k = 0$ for $k \geq 2$. This process is often used to model a semi–Poisson behaviour where the rates of the Poisson arrivals depend on a changing environment.

Example 10.2 Interrupted PH renewal process
Consider a source that emits arrivals according to a PH renewal process. This source may not be working during certain time intervals, due to failure or other reasons of inactivity. Denote the parameters for the second PH distribution (governing the durations of inactivity) (β, U) where U shall be of order n. Denote the number of active states by m. If the source is in an active state i, it may change to inactivity with rate γ_i and to another active state k with rate S_{ik}. Arrivals in state i occur with rate η_i, and the initial distribution of active states is given by the row vector α. Define the off–diagonal entries of the square matrix S to be the rates S_{ik} and set $S_{ii} := -\gamma_i - \sum_{k \neq i} S_{ik} - \eta_i$. Then the interrupted PH renewal process has a BMAP specification

$$D_0 = \begin{pmatrix} S & \gamma\beta \\ \nu\alpha & U \end{pmatrix} \quad \text{and} \quad D_1 = \begin{pmatrix} \eta\alpha & 0 \\ 0 & 0 \end{pmatrix}$$

with $\nu = -U 1_n$, $D_k = 0$ for $k \geq 2$. The name **interrupted PH renewal process** comes from the fact that this process models the behaviour of a source which normally emits arrivals according to a PH renewal process but may be interrupted for PH–distributed periods of time. A classical special case of the interrupted PH renewal process is the interrupted Poisson process (shortly: **IPP**), where $n = m = 1$.

Example 10.3 As we have seen in the beginning of section 9 the time until an overflow occurs in a Markovian queue can be modelled by a PH distribution. Likewise, the output process of Markovian queueing networks with finite buffers can be described as a MAP.

A useful feature of BMAPs is the closure property under superposition. Let $\mathcal{N}_i = (N_t^{(i)} : t \geq 0)$ for $i = 1, 2$ denote two arrival processes. Then the process $\mathcal{N} = (N_t : t \geq 0)$ defined by $N_t := N_t^{(1)} + N_t^{(2)}$ is called the **superposition** of $\mathcal{N}_1$ and $\mathcal{N}_2$. Analogously to the proof of theorem 9.13 one can show

Theorem 10.4 *If $\mathcal{N}_1$ and $\mathcal{N}_2$ are two BMAPs with characterizing matrices $(D_n^{(i)} : n \in \mathbb{N}_0)$ for $i = 1, 2$, then the superposition $\mathcal{N} = \mathcal{N}_1 + \mathcal{N}_2$ is a BMAP with characterizing matrices $D_n = D_n^{(1)} \oplus D_n^{(2)}$ for all $n \in \mathbb{N}_0$.*

4.   Distribution of the Number of Arrivals

The most important random variable of a BMAP is the number N_t of arrivals until some time t. Since BMAPs are Markov processes, we can express all transition probabilities in terms of the generator matrix G and thus in terms of the characterizing matrices D_n. According to theorem 3.7, the transition probability matrix $P(t)$ of a BMAP $\mathcal{Y} = (\mathcal{N}, \mathcal{J})$ with generator G is given by

$$P(t) = e^{G \cdot t} := \sum_{n=0}^{\infty} \frac{t^n}{n!} G^n \tag{10.1}$$

for all $t \geq 0$, with G^n denoting the nth power of the matrix G. Here the entry $P_{k,i;n,j}(t) = \mathbb{P}(N_t = n, J_t = j | N_0 = k, J_0 = i)$ indicates the conditional probability that until time t there have been n arrivals and the phase at time t is j given that at time 0 there already have been k arrivals and the phase is i.

In order to find an expression of $P(t)$ in terms of the matrices D_n, we need to introduce **convolutions of matrix sequences**. Let $\mathcal{M} = (M_n : n \in \mathbb{N}_0)$ and $\mathcal{K} = (K_n : n \in \mathbb{N}_0)$ denote two sequences of $m \times m$ matrices. The convolution $\mathcal{M} * \mathcal{K}$ of these sequences is defined as the sequence $\mathcal{L} = (L_n : n \in \mathbb{N}_0)$ of $m \times m$ matrices with $L_n := \sum_{k=0}^{n} M_k K_{n-k}$ for all $n \in \mathbb{N}_0$.

Define the **convolutional powers of a matrix sequence** $\mathcal{M}$ by the initial sequence $\mathcal{M}^{*0} := (I, 0, 0, \ldots)$ and recursively $\mathcal{M}^{*(n+1)} := \mathcal{M}^{*n} * \mathcal{M}$ for all $n \in \mathbb{N}_0$. If we denote the kth matrix of a sequence $\mathcal{M}^{*n}$ by $\mathcal{M}_k^{*n}$, then we can write $\mathcal{M}_k^{*0} = \delta_{k0} \cdot I$ for all $k \in \mathbb{N}_0$, with δ_{k0} denoting the Kronecker function and I denoting the $m \times m$ identity matrix. Further we have by definition $\mathcal{M}_k^{*1} = M_k$ for all $k \in \mathbb{N}_0$. With these definitions we can state

Lemma 10.5 *For all $n \in \mathbb{N}_0$, the nth power of the generator matrix G of a BMAP has block entries*

$$G_{kl}^n = \Delta_{l-k}^{*n}$$

for all $k \leq l \in \mathbb{N}_0$, *and* $G_{kl}^n = 0$ *for all* $k > l \in \mathbb{N}_0$.

Proof: Fix any $k, l \in \mathbb{N}_0$. Clearly, G^0 is the identity matrix, which proves the statement for $n = 0$. Now assume that the statement is true for some $n \in \mathbb{N}_0$. Then

$$G_{kl}^{n+1} = \sum_{h=0}^{\infty} G_{kh}^n G_{hl} = \sum_{h=k}^{l} G_{kh}^n G_{hl}$$

since $G_{kh}^n = 0$ for $h < k$ by induction hypothesis and $G_{hl} = 0$ for $h > l$ as G is an upper triangular block matrix. In particular we obtain $G_{kl}^{n+1} = 0$ for $k > l$ and we can assume $k \leq l$. The induction hypothesis as well as the structure of G yield $G_{kh}^n = \Delta_{h-k}^{*n}$ and $G_{hl} = D_{l-h}$. Hence we obtain

$$G_{kl}^{n+1} = \sum_{h=k}^{l} \Delta_{h-k}^{*n} D_{l-h} = \sum_{h=0}^{l-k} \Delta_h^{*n} D_{l-k-h} = \Delta_{l-k}^{*(n+1)}$$

which completes the proof.
□

Define the **convolutional exponential of a matrix sequence** $\mathcal{M}$ by the sequence $e^{*\mathcal{M} \cdot t}$ of matrices with the kth matrix being $\sum_{n=0}^{\infty} \frac{t^n}{n!} \mathcal{M}_k^{*n}$ for all $k \in \mathbb{N}_0$. Further define the matrices $P_{kl}(t) = (P_{k,i;l,j}(t))_{i,j \leq m}$. With these definitions and the above lemma we arrive at

Corollary 10.6 *The (k, l)th block entry of the transition probability matrix of a BMAP with characterizing sequence Δ is given by*

$$P_{l-k}(t) := P_{kl}(t) = \left(e^{*\Delta \cdot t}\right)_{l-k} := \sum_{n=0}^{\infty} \frac{t^n}{n!} \Delta_{l-k}^{*n}$$

for all $k \leq l \in \mathbb{N}_0$, *and* $P_{kl}(t) = 0$ *for all* $k > l \in \mathbb{N}_0$.

It is not surprising that we obtain $P_{kl}(t) = 0$ for $k > l$ as a BMAP cannot count more arrivals in any subset of a time interval than in the time interval itself. Furthermore, it is not surprising to see that the block entry $P_{kl}(t)$ depends only on the difference $l - k$, since future arrivals depend only on the current phase and not on the number of arrivals observed in the past. Thus we can abbreviate notations by $P_k(t) := P_{0,k}(t)$. An immediate consequence is

Corollary 10.7 *The matrix containing the probabilities that within a time interval of length t there are no arrivals is given by*

$$P_0(t) = e^{D_0 \cdot t}$$

for all $t \geq 0$.

Define $\tau := \min\{t \geq 0 : N_t > 0)$ as the stopping time until the first arrival. Then the expectation matrix $\mathbb{E}(\tau)$ of τ, having dimension $m \times m$ and entries $\mathbb{E}(\tau \cdot 1_{J_\tau=j}|J_0 = i)$, is given by

$$\mathbb{E}(\tau) = \int_0^\infty e^{D_0 \cdot t}\, dt = -D_0^{-1} \tag{10.2}$$

For a BMAP with characterizing matrices $(D_n : n \in \mathbb{N}_0)$ define the matrix $D := \sum_{n=0}^\infty D_n$. This describes the transition rates of the marginal process $\mathcal{J} = (J_t : t \geq 0)$, which is called the **phase process** of the BMAP. Define the transition probability matrix of J by $P^\Phi(t) = (P_{ij}^\Phi(t))_{i,j \leq m}$ with entries $P_{ij}^\Phi(t) := \mathbb{P}(J_t = j|J_0 = i)$. For this we obtain

Theorem 10.8 *The transition probability matrix of the phase process J is given by*

$$P^\Phi(t) = \sum_{k=0}^\infty P_k(t) = e^{D \cdot t}$$

for all $t \geq 0$ and $i, j \leq m$.

Proof: The first equation holds by definition, since the phase process is the marginal process of $(\mathcal{N}, \mathcal{J})$ in the second dimension. The second equation is obtained via corollary 10.6 and Fubini's theorem as

$$\sum_{k=0}^\infty P_k(t) = \sum_{n=0}^\infty \frac{t^n}{n!} \sum_{k=0}^\infty \Delta_k^{*n} = \sum_{n=0}^\infty \frac{t^n}{n!} \left(\sum_{k=0}^\infty D_k\right)^n = e^{D \cdot t}$$

where the second equality is due to the relation

$$\sum_{k=0}^\infty \Delta_k^{*n} = \left(\sum_{k=0}^\infty D_k\right)^n \tag{10.3}$$

which the reader may prove as an exercise.
□

5.   Expected Number of Arrivals

The expressions $P_k(t)$ will help us to derive a simple representation of the expected number $\mathbb{E}(N_t)$ of arrivals until time t. To this aim we first derive an expression for the z–transform which is defined as $N_t^*(z) := \sum_{n=0}^\infty P_n(t)z^n$

for $z \in \mathbb{C}$ with $|z| \leq 1$. The z–transform of the matrices $(D_n : n \in \mathbb{N}_0)$ is defined as $D(z) := \sum_{n=0}^{\infty} D_n z^n$. Using this, we obtain

Theorem 10.9 *The z–transform of a BMAP having characterizing matrices $(D_n : n \in \mathbb{N}_0)$ is given by*

$$N_t^*(z) = e^{D(z) \cdot t}$$

for all $z \in \mathbb{C}$ with $|z| \leq 1$.

Proof: The definition, together with corollary 10.6, yields

$$N_t^*(z) = \sum_{n=0}^{\infty} \sum_{k=0}^{\infty} \frac{t^k}{k!} \Delta_n^{*k} z^n = \sum_{k=0}^{\infty} \frac{t^k}{k!} \sum_{n=0}^{\infty} \Delta_n^{*k} z^n$$

Since the transform of a k–fold convolution of a sequence equals the kth power of the transform of the sequence (see exercises), we obtain further

$$N_t^*(z) = \sum_{k=0}^{\infty} \frac{t^k}{k!} (D(z))^k = e^{D(z) \cdot t}$$

which is the statement.
□

Let π denote the stationary probability vector of the phase process, satisfying $\pi D = 0$. If the initial phase distribution of a BMAP is π, then we say that the BMAP starts in **phase equilibrium**. Now we can state

Theorem 10.10 *The expected number $\mathbb{E}_\pi(N_t)$ of arrivals until time t, given that the BMAP starts with initial phase distribution π, is determined by*

$$\mathbb{E}_\pi(N_t) = t \cdot \pi \sum_{k=1}^{\infty} k \cdot D_k \mathbf{1}$$

Proof: The first moment can be derived from the z–transform via

$$\mathbb{E}_\pi(N_t) = \pi \frac{d}{dz} \sum_{n=0}^{\infty} \frac{t^n}{n!} (D(z))^n \Bigg|_{z=1} \mathbf{1}$$

$$= \pi \sum_{n=0}^{\infty} \frac{t^n}{n!} \sum_{h=1}^{n} D^{h-1} \sum_{k=1}^{\infty} k \cdot D_k \, D^{n-h} \, \mathbf{1}$$

Now the statement follows from $\pi D = 0$ and $D\mathbf{1} = 0$ which implies that the right–hand sum over h is zero except for $n = h = 1$.
□

The term $\lambda := \pi \sum_{k=1}^{\infty} k D_k \mathbf{1}$ is called the **mean arrival rate** of the BMAP. For the special case of a PH renewal process, this term equals $\pi(\eta\alpha)\mathbf{1} = \pi\eta$. Furthermore the stationary phase distribution π is easily determined via corollary 7.2. Thus we obtain

Corollary 10.11 *For a PH renewal process with parameters (α, T) starting in phase equilibrium, the expected number $\mathbb{E}_\pi(N_t)$ of arrivals until time t is given by*

$$\mathbb{E}_\pi(N_t) = t \cdot \pi\eta$$

with $\eta = -T\mathbf{1}$. The stationary phase distribution π is given by

$$\pi = \frac{1}{-\alpha T^{-1}\mathbf{1}} \int_0^\infty \alpha e^{T \cdot t} \, dt$$

Proof: The first statement is a specification of theorem 10.10. The expression for π is verified by

$$\begin{aligned}
\pi D &= \frac{1}{-\alpha T^{-1}\mathbf{1}} \int_0^\infty \alpha e^{T \cdot t} \, dt \, (T + \eta\alpha) \\
&= \frac{1}{-\alpha T^{-1}\mathbf{1}} \left(\alpha \int_0^\infty e^{T \cdot t} T \, dt + \int_0^\infty \alpha e^{T \cdot t} \eta \, dt \, \alpha \right) \\
&= \frac{1}{-\alpha T^{-1}\mathbf{1}} (\alpha(-I) + \alpha) = 0
\end{aligned}$$

with I denoting the identity matrix, and

$$\int_0^\infty \alpha e^{T \cdot t} \, dt \, \mathbf{1} = \alpha \int_0^\infty e^{T \cdot t} \, dt \, \mathbf{1} = -\alpha T^{-1}\mathbf{1}$$

□

The above PH renewal process $\mathcal{N}$ is a delayed renewal process with initial delay $X_0 \sim PH(\pi, T)$ and renewal intervals $X_n \sim PH(\alpha, T)$ for $n \in \mathbb{N}$. The elementary renewal theorem 6.12 then states that $\lim_{t\to\infty} \mathbb{E}(N_t)/t = 1/\mathbb{E}(X_1)$. Thus we obtain another expression for the mean of a phase–type distributed random variable (cf. corollary 9.9).

Corollary 10.12 *For a $PH(\alpha, T)$ distributed random variable X the expectation is given by $\mathbb{E}(X) = (\pi\eta)^{-1}$.*

Notes

The first presentation of the PH renewal process has been given in Neuts [63]. The MAP in the present formulation has been introduced first in Lucantoni et al. [56] as a generalization of the PH renewal process and the Markov–modulated Poisson process. Its further generalization to the BMAP with batch arrivals has been introduced in Lucantoni [54]. An early algorithms for a computation of the transition probabilities can be found in Neuts and Li [67]. The calculus of matrix convolutions that leads to explicit expressions for the transition probabilities of BMAPs has been introduced in Baum [9] or Breuer [21].

The class of BMAPs is equivalent to the class of **versatile Markovian point processes** (or **N–process**es) introduced in Neuts [64]. However, this formulation is more complicated and does not yield as simple an anlysis of the respective queueing systems. For generalizations of BMAPs see Pacheco and Prabhu [68], Baum and Kalashnikov [11, 13], or Breuer [19, 21]. A result which is analogous to theorem 9.14 is that the class of all MAPs is dense within the class of all marked point processes (see Asmussen and Koole [6]).

An extensive treatment on the history of the BMAP, along with many references, can be found in Lucantoni [55]. Statistical methods for fitting MMPP and BMAP models are given in Ryden [76] and Breuer [18].

Exercise 10.1 Prove theorem 10.4.

Exercise 10.2 Prove equation (10.3) as well as

$$\sum_{n=0}^{\infty} \Delta_n^{*k} z^n = (D(z))^k$$

for $z \in \mathbb{C}$ with $|z| < 1$.

Exercise 10.3 Give a model (in terms of a PH renewal process) for the number of orders the plumber of exercise 6.4 receives.

Exercise 10.4 Show that an IPP is uniquely determined by the first three moments of the inter–arrival time distribution.

Exercise 10.5 In a telephone network data transmission via the package voice system works as follows. The language source is digitalized and divided into packages, which are to be transmitted. A language source switches between "talk spurts" and "silent periods". Thus a model in terms of an IPP seems reasonable.

Measurements yield a mean inter–arrival time of 3 ms for the packages. The squared coefficient of variation is 300. Further it is known that a silent period is about two times as long as a talk spurt. Adjust the IPP to these measurements.

Exercise 10.6 Describe the BMAP/PH/k queue in terms of a Markov process with $(k + 2)$–dimensional state space.

Chapter 11

THE GI/PH/1 QUEUE

In section 1, we have analyzed one of the classical semi–Markovian queueing systems, the GI/M/1 queue. For practical applications, this model has the disadvantage that the assumption of exponential service times is often inappropriate for the actual service time distribution governing the system. More typical service times often are distributed like generalized Erlang or Cox distributions, or special distributions like the lognormal or Weibull type. The former are special cases of phase–type distributions, the latter can be approximated by them. Thus the wish to extend the results for the GI/M/1 queue to GI/PH/1 queues is understandable.

It will turn out in the presentation of this chapter that the analysis of GI/PH/1 queues can be performed in a strikingly similar manner to the GI/M/1 analysis. This is one of the main reasons for the success of the phase–type concept in queueing theory.

The **GI/PH/1 queue** is characterized by the following features. Arrivals are generated by a renewal process with inter–arrival times distributed by a distribution function H with $0 < \mathbb{E}(H) < \infty$. The service times are iid according to a $PH(\alpha, T)$ distribution of order $m \in \mathbb{N}$, with $\alpha_{m+1} = 0$. There is one server, and the service discipline is FCFS. The waiting room capacity is infinite such that there are no users lost.

In order to keep a complete description of the system state, it is not sufficient anymore to remember the number of users in the system only, but we need to keep track of the current phase of the service time distribution, too. Thus we examine the process $\mathcal{Q} = (Q_t : t \geq 0)$ with $Q_t = (N_t, J_t)$ for all $t \geq 0$, where N_t and J_t denote the number of users in the system and the current phase of service at time t, respectively. If the system is empty, i.e. $N_t = 0$, there is no service and we do not need to keep track of a phase of service. In this case we set $J_t = 0$. Thus the state space of $\mathcal{Q}$ is $E = \{0, 0\} \cup \mathbb{N} \times \{1, \ldots, m\}$.

1.   The Embedded Markov Chain

As for the GI/M/1 queue we observe that at times of arrivals the conditions for a prognosis of the future of the system are less difficult, since we know that the time until the next arrival is distributed by F. This is due to the iid inter–arrival times and the independence of the arrival process from the rest of the system. Hence we can construct a Markov chain embedded at arrival instants.

Define T_n as the time of the nth arrival and $X_n := (N_{T_n} - 1, J_{T_n})$ for all $n \in \mathbb{N}$. Since at times T_n there is always at least one user in the system (namely the one that has just arrived), the state space of $\mathcal{X} = (X_n : n \in \mathbb{N})$ is $\mathbb{N}_0 \times \{1, \ldots, m\}$. If we know that the phase of service currently is j, then we also know that the time until the next service completion will be distributed by a $PH(e_j, T)$ distribution, with e_j denoting the jth canonical row base vector. Since we further know that the time until the next arrival after T_n will be distributed by F (the inter–arrival time has just begun), the chain $\mathcal{X}$ is Markovian.

Now we want to determine the transition matrix $\tilde{P}$ of $\mathcal{X}$. Since the state space of $\mathcal{X}$ is two–dimensional, the structure of $\tilde{P}$ must become more complicated than for the analogue of the GI/M/1 queue. However, we can order the state space of $\mathcal{X}$ lexicographically as

$$\{(0, 1), \ldots, (0, m), (1, 1), \ldots, (1, m), (2, 1), \ldots\}$$

such that the transition matrix $\tilde{P}$ will have a block structure. The first dimension of the state space shall be called the **level** of the process, while the second dimension is called the **phase**. The chain $pr_2(\mathcal{X})$, with pr_2 denoting the projection on the second dimension, will be called the (embedded) **phase process**.

The general structure of $\tilde{P}$ is determined by two considerations. First, between two arrival instants the number of users in the system can increase by at most one. Second, as long as the system is not empty, the change of the number of users between two arrival instants T_n and T_{n+1} does not depend on the number of users in the system at time T_n. Thus we obtain the structure

$$\tilde{P} = \begin{pmatrix} B_0 & A_0 & & & \\ B_1 & A_1 & A_0 & & \\ B_2 & A_2 & A_1 & A_0 & \\ \vdots & \vdots & \ddots & \ddots & \ddots \end{pmatrix} \tag{11.1}$$

for the transition matrix $\tilde{P}$, with the non–specified entries being zero matrices. We see that the matrix $\tilde{P}$ has a **block Toeplitz structure** with one upper diagonal and special boundary entries in the first column. In terms of levels, the matrix is **skip–free to the right** as in the GI/M/1 case. Because of the similarity to the GI/M/1 case, such a matrix is also called a **GI/M/1 type matrix**.

Here, the (i, j)th entry $A_{k;ij}$ of the matrix A_k indicates the probability that between two consecutive arrivals instants T_n and T_{n+1}, k users are served and the phase of service changes from i at time T_n to j at time T_{n+1}. The entries $B_{k;ij}$ indicate the probabilities that between two consecutive arrivals instants T_n and T_{n+1}, at least $k + 1$ users are served and the phase of service changes from i at time T_n to j at time T_{n+1}.

In order to find expressions for the matrices A_k and B_k, we define the following family of matrices. For every $t \geq 0$, $k \in \mathbb{N}_0$, and $i, j \in \{1, \ldots, m\}$ let $P_{k;ij}(t)$ denote the probability that during the interval $[0, t]$ there are k service completions and the phase of service at time t is j given that the phase of service at time 0 has been i. Further define the respective $m \times m$ matrices $P_k(t) = (P_{k;ij}(t))_{i,j \leq m}$. Then we can write

$$A_k = \int_0^\infty P_k(t) \, dH(t) \quad \text{and} \quad B_k = \sum_{i=k+1}^\infty A_i$$

for all $k \in \mathbb{N}_0$. The matrices $P_k(t)$ can be determined according to the expression given in corollary 10.6 by specifying $\Delta = (T, \eta\alpha, 0, 0, \ldots)$, with $\eta = -T\mathbf{1}$. Since $A := \sum_{n=0}^\infty A_n$ is stochastic, we immediately obtain

$$B_k = \mathbf{1}\alpha - \sum_{i=0}^k A_i \mathbf{1}\alpha$$

for all $k \in \mathbb{N}_0$.

2. Stationary Distribution at Arrival Instants

From now on we shall assume that the **stability condition** $\mathbb{E}(H) > -\alpha T^{-1}\mathbf{1}$ for the embedded Markov chain $\mathcal{X}$ is satisfied. In the next section it is shown that then a stationary distribution for $\mathcal{X}$ does exist. In this section we want to show how this can be determined.

Denote the stationary distribution for $\mathcal{X}$ by $\mathbf{x} = \mathbf{x}\tilde{P}$ and write

$$\mathbf{x} = (\mathbf{x}_n : n \in \mathbb{N}_0) = (x_{ni} : n \in \mathbb{N}_0, 1 \leq i \leq m)$$

with x_{ni} being the stationary probability that the chain $\mathcal{X}$ is in state (n, i). The vectors $\mathbf{x}_n$ contain the stationary probabilities of the chain $\mathcal{X}$ being in level n.

For the GI/M/1 queue we could derive a geometric structure $\nu_{n+1} = \nu_n \xi$ for the stationary distribution ν of the embedded Markov chain. In the present

case of a GI/PH/1 queue we will have a similar structure for the distribution
$\mathbf{x}$, namely the relation $\mathbf{x}_{n+1} = \mathbf{x}_n R$ where now the factor R is a matrix.
To underline the analogy, such a distribution is called a **matrix–geometric
distribution** and R is called **rate matrix**.

We will first define the matrix R and then show that $\mathbf{x}$ indeed is matrix–
geometric with rate matrix R. For all phases $i, j \leq m$ and levels $k \geq 0$,
define the **taboo probabilities** (cf. definition (8.11))

$$_k P^{(n)}_{k,i;k+1,j} := \mathbb{P}(X_n = (k+1,j), pr_1(X_h) \neq k \ \forall \ 1 \leq h < n | X_0 = (k,i))$$

that the chain $\mathcal{X}$ enters level $k + 1$ in phase j after n steps and does not enter
level k before, given that it starts in level k and phase i.

The structure of $\tilde{P}$ is self–similar in the sense that if we delete the first n
(block) rows and columns of $\tilde{P}$ for any $n \in \mathbb{N}$, we obtain the same matrix as
if we deleted only the first row and column. This implies that the probabilities
$_k P^{(n)}_{k,i;k+1,j}$ are independent of $k \in \mathbb{N}_0$. Hence we can define

$$r_{ij} := \sum_{n=1}^{\infty} {}_k P^{(n)}_{k,i;k+1,j}$$

for all $i, j \in \{1, \ldots, m\}$, independently of $k \geq 0$. The value r_{ij} is the expected
number of visits of the chain $\mathcal{X}$ to the state $(k + 1, j)$ before returning to level
k if $\mathcal{X}$ is started in state (k, i). Finally, we define the matrix $R = (r_{ij})_{i,j \leq m}$
and call it the rate matrix.

Theorem 11.1 *The stationary distribution* $\mathbf{x} = (\mathbf{x}_k : k \in \mathbb{N}_0)$ *of* $\mathcal{X}$ *satisfies
the relation*

$$\mathbf{x}_{k+1} = \mathbf{x}_k R$$

for all $k \in \mathbb{N}_0$.

Proof: Fix any level $k \in \mathbb{N}_0$. By conditioning on the last time and phase of
the last visit to level k we obtain the relation

$$P^{(n)}_{k+1,j;k+1,j} = {}_k P^{(n)}_{k+1,i;k+1,j} + \sum_{i=1}^{m} \sum_{r=1}^{n} P^{(r)}_{k+1,j;k,i} \cdot {}_k P^{(n-r)}_{k,i;k+1,j}$$

for all $n \geq 1$.

If we add these equations for $n = 1, \ldots, N$, divide both sides by N and then
let N tend to infinity, the left–hand side tends to $x_{k+1,j}$ according to corollary

2.28. Since the sum $\sum_{n=1}^{\infty} {}_k P_{k+1,i;k+1,j}^{(n)}$ is finite if $\mathcal{X}$ is positive recurrent, the first term on the right–hand side tends to zero. The second term equals

$$\lim_{N\to\infty} \sum_{i=1}^m \frac{1}{N} \sum_{n=1}^N \sum_{r=1}^n P_{k+1,j;k,i}^{(r)} \cdot {}_k P_{k,i;k+1,j}^{(n-r)}$$

$$= \lim_{N\to\infty} \sum_{i=1}^m \frac{1}{N} \sum_{r=1}^N P_{k+1,j;k,i}^{(r)} \cdot \sum_{n=1}^{N-r} {}_k P_{k,i;k+1,j}^{(n)}$$

$$= \sum_{i=1}^m \lim_{N\to\infty} \frac{1}{N} \sum_{r=1}^N P_{k+1,j;k,i}^{(r)} \cdot \lim_{N\to\infty} \sum_{n=1}^{N-r} {}_k P_{k,i;k+1,j}^{(n)} = \sum_{i=1}^m x_{ki} \cdot r_{ij}$$

by corollary 2.28 and the definition of R.
□

The matrix R contains the expected number of visits to any level $l + 1 \in \mathbb{N}$ between two consecutive visits to level l. In case of positive recurrence this matrix is entry–wise finite by theorem 2.24. For the powers of R we obtain the following interpretation:

Theorem 11.2 *For any $k \in \mathbb{N}$, the (i, j)th entry of the matrix R^k indicates the expected number of visits to the state $(l + k, j)$ between two consecutive visits to level $l \in \mathbb{N}_0$, given that the chain $\mathcal{X}$ starts in state (l, i).*

Proof: The statement holds for $k = 1$ by definition of R. Now we assume that it holds for some $k \in \mathbb{N}$ and want to show the induction step to $k + 1$. For $n \geq k + 1$ the relation

$$ {}_l P_{l,i;l+k+1,j}^{(n)} = \sum_{h=1}^m \sum_{r=0}^n {}_l P_{l,i;l+k,h}^{(r)} \cdot {}_{l+k} P_{l+k,h;l+k+1,j}^{(n-r)}$$

$$= \sum_{h=1}^m \sum_{r=0}^n {}_l P_{l,i;l+k,h}^{(r)} \cdot {}_l P_{l,h;l+1,j}^{(n-r)}$$

is obtained after conditioning on the time r and the phase h of the last visit to level $l + k$ before visiting level $l + k + 1$. For $n \leq k$ both sides of the equation are zero, such that we can sum over all $n \geq 0$. The left–hand side sums up to the desired expectation for the level $k + 1$. On the right–hand side we obtain

$$\sum_{h=1}^m \sum_{n=0}^\infty \sum_{r=0}^n {}_l P_{l,i;l+k,h}^{(r)} \cdot {}_l P_{l,h;l+1,j}^{(n-r)} = \sum_{h=1}^m \sum_{r=0}^\infty {}_l P_{l,i;l+k,h}^{(r)} \sum_{n=0}^\infty {}_l P_{l,h;l+1,j}^{(n)}$$

$$= \sum_{h=1}^m R_{ih}^k r_{hj}$$

which completes the induction step.

□

There is an interesting relation between the matrix R and its powers, which at
the same time yields a method to compute R without needing to calculate the
n–step taboo probability matrices $_lP^{(n)}$ by which R is defined. This is given
in

Theorem 11.3 *If $\mathcal{X}$ is positive recurrent, then the matrix R is the (entry–wise)
minimal non–negative solution to the matrix equation*

$$M = \sum_{n=0}^{\infty} M^n A_n$$

*It can be obtained as the limit $R = \lim_{k \to \infty} M(k)$ with $M(0) := 0$ and
$M(k+1) := \sum_{n=0}^{\infty} M(k)^n A_n$ for all $k \in \mathbb{N}_0$.*

Proof: First we observe $_lP^{(1)}_{l,i;l+1,j} = A_{0;ij}$. For $n \geq 2$ we obtain

$$_lP^{(n)}_{l,i;l+1,j} = \sum_{h=1}^{m} \sum_{k=1}^{\infty} {}_lP^{(n-1)}_{l,i;l+k,h} \cdot A_{k;hj}$$

by conditioning on the state $(l+k, h)$ from which the level l is entered for time
n. If we sum up these equations for all n, we obtain $R = \sum_{n=0}^{\infty} R^n A_n$.

For the second statement, consider the sequence $M = (M(k) : k \in \mathbb{N}_0)$.
Clearly, $M(1) \geq M(0)$ and $R \geq M(0)$ entry–wise. The fact that

$$M(k+1) - M(k) = \sum_{n=0}^{\infty} (M(k)^n - M(k-1)^n) A_n \geq 0$$

$$R - M(k) = \sum_{n=0}^{\infty} (R^n - M(k-1)^n) A_n \geq 0$$

yields by induction that the sequence M is entry–wise monotonically increas-
ing and $M(k) \leq R$ for all $k \in \mathbb{N}$. Hence there is a matrix $M^* = \lim_{k \to \infty} M(k)$
with $M^* \leq R$. Thus we can use the dominated convergence theorem to verify

$$M^* = \lim_{k \to \infty} M(k) = \lim_{k \to \infty} \sum_{n=0}^{\infty} M(k-1)^n A_n = \sum_{n=0}^{\infty} \lim_{k \to \infty} M(k-1)^n A_n$$

$$= \sum_{n=0}^{\infty} (M^*)^n A_n$$

such that M^* is indeed a solution to the matrix equation.

Every other non–negative solution M' to the matrix equation must satisfy $M' \geq A_0 = M(1)$. But $M' \geq M(k)$ implies

$$M' = \sum_{n=0}^{\infty} (M')^n A_n \geq M(k)^n A_n = M(k+1)$$

and thus by induction $M' \geq M(k)$ for all $k \in \mathbb{N}$ and in the limit $M' \geq M^*$. Hence M^* is minimal.

It remains to show that $R \leq M^*$. Define the matrices

$$_l P_r^{(n)} := \left(_l P_{l,i;l+r,j}^{(n)} \right)_{i,j \leq m}$$

for all $n, r \in \mathbb{N}_0$ and remember that their definition is independent of the start level l. The matrix $_l P_r^{(n)}$ contains the (phase–dependent) probabilities that the level of the chain is raised by r after exactly n steps. Because of the structure (11.1) it is clear that $_l P_r^{(n)} = 0$ for $r > n$.

Further define the matrices $Y_s^{(r)} := \sum_{n=0}^{s} {}_l P_r^{(n)}$ and set $Y_s := Y_s^{(1)}$. Then we can write $R = \lim_{s \to \infty} Y_s$. The matrix $Y_s^{(r)}$ contains the (phase–dependent) expectations of the number of visits to the level $l + r$ within the course of s steps beginning from taboo level l. Again the structure (11.1) yields $Y_s^{(r)} = 0$ for $r > s$.

Because of $_l P_1^{(n)} = \sum_{r=0}^{l-1} {}_l P_r^{(n-1)} A_r$ we obtain

$$Y_s = \sum_{n=1}^{s} {}_l P_1^{(n)} = \sum_{k=0}^{s-1} \sum_{r=0}^{k} {}_l P_r^{(k)} A_r = \sum_{r=0}^{\infty} \left(\sum_{k=0}^{s-1} {}_l P_r^{(k)} \right) A_r = \sum_{r=0}^{\infty} Y_{s-1}^{(r)} A_r$$

for all $s \in \mathbb{N}$. Conditioning on the time k of the last visit to level $l + r - 1$ yields

$$_l P_r^{(n)} = \sum_{k=0}^{n} {}_l P_{r-1}^{(k)} \, {}_l P_1^{(n-k)}$$

which implies

$$Y_{s-1}^{(r)} = \sum_{n=0}^{s-1} \sum_{k=0}^{n} {}_l P_{r-1}^{(k)} \, {}_l P_1^{(n-k)} \leq \sum_{k=0}^{s-1} {}_l P_{r-1}^{(k)} \sum_{n=0}^{s-1} {}_l P_1^{(n)} = Y_{s-1}^{(r-1)} Y_{s-1}^{(1)}$$

whence $Y_{s-1}^{(r)} \leq Y_{s-1}^r$ for all $r \geq 1$. Thus Y_s satisfies the inequality

$$Y_s \leq \sum_{n=0}^{\infty} Y_{s-1}^n A_n$$

Clearly, the sequence $(Y_s : s \in \mathbb{N})$ is increasing and $Y_1 = A_0 = M(1)$, while $Y_2 \leq M(2)$. Now the above inequality yields $Y_s \leq M(s)$ for all $s \in \mathbb{N}$, which implies the desired bound $R \leq M^*$.
$\square$

Theorem 11.4 *The stationary probability vector x_0 of $\mathcal{X}$ is determined by*

$$x_0 = x_0 B[R] \qquad \text{and} \qquad x_0 (I - R)^{-1} \mathbf{1} = 1$$

with $B[R] = \sum_{n=0}^{\infty} R^n B_n$.

Proof: The stationarity of $x = x\tilde{P}$ yields according to the structure given in (11.1) and theorem 11.1

$$x_0 = \sum_{n=0}^{\infty} x_n B_n = x_0 \sum_{n=0}^{\infty} R^n B_n$$

which proves the first equation. In order that $x = (x_k : k \in \mathbb{N}_0)$ be a probability distribution, we further need

$$1 = \sum_{k=0}^{\infty} x_k \mathbf{1} = x_0 \sum_{k=0}^{\infty} R^k \mathbf{1} = x_0 (I - R)^{-1} \mathbf{1}$$

which is the second equation. Since $\tilde{P}$ is irreducible, there is at most one stationary distribution of $\mathcal{X}$. Since the two equations in the statement and the relation in theorem 11.1 yield a solution to $x = x\tilde{P}$, the vector x_0 is uniquely determined.
$\square$

After verifying that $B[R] = \mathbf{1}\alpha$ (as an exercise), we immediately obtain

Corollary 11.5 *The stationary probability vector x_0 of $\mathcal{X}$ is explicitly given as*

$$x_0 = (\alpha(I - R)^{-1}\mathbf{1})^{-1}\alpha$$

3.   Ergodicity of the Embedded Markov Chain

In the previous section we have derived the form of the stationary distribution in case of positive recurrence of the embedded Markov chain $\mathcal{X}$. Now we want to derive the conditions for positive recurrence, also called the **ergodicity conditions** for $\mathcal{X}$.

First a remark is due regarding irreducibility of $\mathcal{X}$. If $\alpha_i = 0$ for some phase $1 \leq i \leq m$, then for all matrices B_k the ith column vanishes, and $\mathcal{X}$ is not irreducible. However, all statements concerning existence and uniqueness of a stationary distribution hold as if the chain $\mathcal{X}$ were irreducible. This is due to remark 2.26, which was to be proven as an exercise in chapter 2.

Since in case of positive recurrence the structure of the stationary distribution **x** is largely determined by R, it seems natural to search for some property of R to yield the ergodicity condition. This can be found in the second equation in theorem 11.4.

Theorem 11.6 *The embedded Markov chain* $\mathcal{X}$ *is positive recurrent if and only if* $(I - R)$ *is invertible.*

Proof: Define the vector $\mathbf{v} = \sum_{n=0}^{\infty} R^n \mathbf{1}$, which may have infinite entries. According to theorem 11.2 the value v_i indicates the expected return time to level zero given that $X_0 = (0, i)$. By theorem 2.32 (with F being the states of the level zero), the chain $\mathcal{X}$ is positive recurrent if and only if $\mathbf{v}$ is entry–wise finite. This is equivalent to convergence of the series $\sum_{n=0}^{\infty} R^n = (I - R)^{-1}$. $\square$

Although the above stability condition is the first that comes into mind if we regard the structure of the stationary distribution **x**, it may be difficult to check this condition if there are eigenvalues of R with modulus close to one. Seen from a more general perspective, we would expect the queue to be stable if and only if the classical condition, namely that the mean service time be less than the mean inter–arrival time, holds. That this is indeed the case can be shown for the $GI/PH/1$ queue, too. However, to this aim we first need to translate the classical condition to an equivalent condition in terms of the system parameters.

We have denoted the distribution function of the inter–arrival times by H and its mean by $\mathbb{E}(H)$. According to corollary 9.9, the mean service time is given by $\mathbb{E}(S) = -\alpha T^{-1} \mathbf{1}$. Then the classical stability condition may be stated as

$$\rho = \frac{-\alpha T^{-1} \mathbf{1}}{\mathbb{E}(H)} < 1$$

Now define the matrix $A := \sum_{n=0}^{\infty} A_n$, which contains the transition probabilities for the phase process at times of arrivals, given that the server is busy. Clearly A is stochastic. Note further that A is irreducible, due to the phase–type service process without superfluous phases. Let π denote the stationary vector for A, satisfying $\pi A = \pi$. If we define further $D := T + \eta \alpha$, then we can state

Lemma 11.7 *The vector* π *satisfies* $\pi D = 0$.

Proof: By definition of A and theorem 10.8 we obtain

$$\pi = \pi \sum_{n=0}^{\infty} A_n = \pi \int_0^{\infty} \sum_{n=0}^{\infty} P_n(t) \, dH(t) = \pi \int_0^{\infty} e^{D \cdot t} \, dH(t)$$

$$= \int_0^{\infty} \pi \sum_{n=0}^{\infty} \frac{t^n}{n!} D^n \, dH(t)$$

Clearly this equation holds if $\pi D = 0$. Since the matrix A is irreducible and therefore the vector π uniquely determined, the statement follows.
□

Theorem 11.8 *The classical stability condition $\rho < 1$ is equivalent to the condition*

$$\pi \sum_{k=1}^{\infty} k \cdot A_k \, \mathbf{1} > 1$$

Proof: The definition of the A_k yields

$$\pi \sum_{k=1}^{\infty} k \cdot A_k \, \mathbf{1} = \int_0^{\infty} \pi \sum_{k=1}^{\infty} k \cdot P_k(t) \, \mathbf{1} \, dH(t)$$

Here the expression within the integral is the expected number of service completions in time t given that the service process is started in phase distribution π. Since the sequence of service completions under the regime of a busy server is a PH renewal process, corollary 10.11 yields

$$\pi \sum_{k=1}^{\infty} k \cdot A_k \, \mathbf{1} = \int_0^{\infty} \mathbb{E}_\pi(N_t) \, dH(t) = \pi \eta \cdot \mathbb{E}(H)$$

Now the expression for π in corollary 10.11 implies

$$\pi \eta = - \left(\alpha T^{-1} \mathbf{1} \right)^{-1} \int_0^{\infty} \alpha e^{T \cdot t} \eta \, dt = - \left(\alpha T^{-1} \mathbf{1} \right)^{-1}$$

Thus we obtain

$$\pi \sum_{k=1}^{\infty} k \cdot A_k \, \mathbf{1} = \frac{1}{\rho} \tag{11.2}$$

which completes the proof.
□

The above condition postulates that under the stationary phase regime the mean number of service completions between two arrivals is greater than one. It is as intuitive as the classical stability condition and can be verified almost immediately once the system parameters are given. Now we will show that this condition implies positive recurrence of the embedded Markov chain $\mathcal{X}$.

We write $A = \sum_{n=0}^{\infty} A_n$ as usual, and denote by $\bar{\mathbf{a}} = (\bar{a}_1, \ldots, \bar{a}_m)$ the column vector of components $\bar{a}_i = \sum_{j \in E} \sum_{n=0}^{\infty} n A_{n;ij}$ for $1 \leq i \leq m$. In order to apply the criteria of theorem 2.33 to the chain $\mathcal{X}$ we first construct a non-negative solution $\mathbf{v}$ of

$$(\mathbf{I} - A)\mathbf{v} = \bar{\mathbf{a}} - (1/\rho)\mathbf{1} \tag{11.3}$$

Lemma 11.9 *The system (11.3) possesses solutions of the form $\mathbf{v} + r\mathbf{1}$ with $r \in \mathbb{R}$, where $\mathbf{v}$ is any finite separate solution and $\{r\mathbf{1} : r \in \mathbb{R}\}$ represents the set of all solutions of the homogeneous system. In particular, it is always possible to find a solution $\mathbf{v}$ that is non-negative.*

Proof: Since A is non-negative and stochastic, $A - \mathbf{I}$ represents the generator of some Markov process and, consequently, has rank $m - 1$. The stationary vector of that process is π (since $\pi A = \pi$).

An inhomogeneous system $M\mathbf{v} = \mathbf{d}$ of m linear equations with m unknowns, whose characteristic matrix M is of rank $k < m$, possesses a solution if and only if the vector of the right–hand side is orthogonal to all solution vectors $\mathbf{w} = (w_1, \ldots, w_m)$ of the adjoint homogeneous system $\mathbf{w}M = \mathbf{0}$. In our case the rank is $m - 1$, and any solution of the adjoint system is some multiple of π. Checking the condition of orthogonality, we see that

$$\pi\,(\bar{\mathbf{a}} - (1/\rho)\mathbf{1}) = \pi \sum_{n=0}^{\infty} n\,A_n \mathbf{1} - \frac{1}{\rho} = 0$$

because of (11.2). Therefore, the system (11.3) possesses a finite solution $\mathbf{v} = (v_1, \ldots, v_m)$. The general solution is the sum of a separate solution and a solution of the homogeneous system. Since, obviously, the homogeneous system has as solutions all multiples of $\mathbf{1} = (1, 1, \ldots, 1)^T$, the general non–homogeneous solution is $\mathbf{v} + r \cdot \mathbf{1}$, for arbitrary $r \in \mathbb{R}$. $\square$

Theorem 11.10 *The embedded Markov chain $\mathcal{X}$ is positive recurrent if $\rho < 1$.*

Proof: Applying Lemma 11.9, let $\mathbf{v} = (v_1, \ldots, v_m)$ be some non-negative bounded solution of equation (11.3). Using this solution, define the finite function

$$f(s,j) = \begin{cases} s + v_j & \text{for } s \in \mathbb{N}, \ 1 \leq j \leq m \\ 0 & \text{for } s = 0, \ 1 \leq j \leq m \end{cases}, \tag{11.4}$$

such that for $r > 0$,

$$\sum_{(s,j)} \tilde{P}_{r,i;s,j} f(s,j) - f(r,i) = \sum_{(s,j)} \tilde{P}_{r,i;s,j} \cdot (s - r) + \sum_{s \geq 1; j \leq m} \tilde{P}_{r,i;s,j} v_j - v_i$$

$$= -r \sum_{j=1}^{m} B_{r;ij} + \sum_{j=1}^{m} \sum_{n=0}^{r} A_{n;ij} \cdot (1 - n) + \sum_{j=1}^{m} \sum_{n=0}^{r} A_{n;ij} v_j - v_i$$

Since $\sum_{j=1}^{m} \sum_{n=0}^{\infty} n A_{n;ij} < \infty$, the sums $\sum_{j=1}^{m} \sum_{n=r}^{\infty} n A_{n;ij}$ tend to zero as r tends to infinity. Hence

$$r \sum_{j=1}^{m} B_{r;ij} = \sum_{j=1}^{m} \sum_{n=r+1}^{\infty} r A_{n;ij} \to 0 \quad \text{as} \quad r \to \infty$$

If $\rho < 1$, then equation (11.3) yields

$$\lim_{r \to \infty} \left(\sum_{(s,j)} \tilde{P}_{r,i;s,j} f(s,j) - f(r,i) \right) \leq 1 - \sum_{j=1}^{m} \sum_{n=0}^{\infty} n A_{n;ij} + \bar{a}_i - \frac{1}{\rho}$$

$$= 1 - \frac{1}{\rho} < 0,$$

implying that there exist an $r_0 \in \mathbb{N}$ and an $\varepsilon > 0$, such that for $\rho < 1$

$$L(r,i) = \sum_{(s,j)} \tilde{P}_{r,i;s,j} \cdot f(s,j) - f(r,i) < -\varepsilon \quad \text{for all } r \geq r_0, \, i \leq m.$$

Define a finite subset $F \subset \mathbb{N}_0 \times E$ of states as

$$F = \{(r,i) : r < r_0, i \leq m\}. \tag{11.5}$$

Since for $r = 0$ the expression $\sum_{j=1}^{m} A_{0;ij}(1 + x_j)$ is positive (bearing in mind that $v_j \geq 0$ for all j), this set F is not empty. Since $L(r,i) \leq -\varepsilon$ for $(r,i) \notin F$, and $0 \leq f(s,j) < \infty$ for all $s \in \mathbb{N}_0$ and $1 \leq j \leq m$, the prerequisites of theorem 2.33 are satisfied, i. e. the chain is positive recurrent. $\square$

4.   Asymptotic Distribution of the System Process

If the stability condition $\rho < 1$ is satisfied, we have seen in the previous two sections how to derive the stationary distribution **x** of the embedded Markov chain at arrival instants. Now we want to use this in order to obtain an expression for the asymptotic distribution of the system process $\mathcal{Q} = (Q_t : t \geq 0)$. Here $Q_t = (n, j)$ states that there are n users in the system (including the server) and the phase of the service is j at time t. If there are no users in the system, there is no service either. This is denoted by $Q_t = (0,0)$.

To obtain the asymptotic distribution of $\mathcal{Q}$ is a simple matter if we conceive the system process as a semi-regenerative process in the same way as we already did in the analysis of the GI/M/1 queue. If $\mathcal{T} = (T_n : n \in \mathbb{N})$ denotes the sequence of arrival instants and $\mathcal{X} = (X_n : n \in \mathbb{N})$ is defined as in section 1, then the independence of the arrival process from the rest of the system yields

$$G_{(k,i),(n,j)}(t) = \mathbb{P}(T_{n+1} - T_n \leq t | X_n = (k,i), X_{n+1} = (n,j)) = H(t)$$

independently of $k, n \in \mathbb{N}_0$ and $i, j \leq m$. The standard assumption $\mathbb{E}(H) > 0$ on the distribution function H implies that $T_n \to \infty$ $\mathbb{P}$–almost certainly as $n \to \infty$. Hence $(\mathcal{X}, \mathcal{T})$ is a Markov renewal chain and $\mathcal{Q}$ is semi–regenerative.

In order to employ theorem 7.15 for the calculation of the asymptotic distribution of the system process $\mathcal{Q}$, we already have determined the stationary distribution $\nu = \mathbf{x}$ of $\mathcal{X}$ in section 2. It remains to derive the vector $\mathbf{m}$ of the mean time between Markov renewal points as well as the function $K(t)$ describing the behaviour of the system process between Markov renewal points. The vector $\mathbf{m}$ is obtained in a straightforward manner as

$$m_{n,i} = \mathbb{E}(T_1 | X_0 = (n,i)) = \mathbb{E}(H) \tag{11.6}$$

independently of $n \in \mathbb{N}_0$ and $i \leq m$, since the arrival process does not depend on the state of the system. Thus the vector $\mathbf{m}$ is constant. The function $K(t)$ is given by the values $K_{(k,i),(n,j)}(t) = \mathbb{P}(T_1 > t, Q_t = (n,j) | X_0 = (k,i))$. Exploiting the independence of arrival process and service and abbreviating $H^c(t) := 1 - H(t)$, we obtain

$$K_{(k,i),(n,j)}(t) = \begin{cases} H^c(t) \cdot P_{k+1-n;ij}(t), & 1 \leq n \leq k+1 \\ H^c(t) \cdot \sum_{h=k+1}^{\infty} \sum_{j=1}^{m} P_{h;ij}(t), & (n,j) = (0,0) \\ 0, & n > k+1 \end{cases}$$

for all $t > 0$, $k, n \in \mathbb{N}_0$, and $i, j \leq m$.

Denote the asymptotic distribution of the system process by the block vector $\mathbf{y} = (\mathbf{y}_n : n \in \mathbb{N}_0)$. For $n \in \mathbb{N}$ the blocks $\mathbf{y}_n = (y_{ni} : 1 \leq i \leq m)$ are defined by $y_{ni} = \lim_{t \to \infty} \mathbb{P}(Q_t = (n,i))$. The asymptotic probability of the system being idle is denoted by $\mathbf{y}_0 = y_{00} = \lim_{t \to \infty} \mathbb{P}(Q_t = (0,0))$. Now theorem 7.15 yields

Corollary 11.11 *If the stability condition $\rho < 1$ holds, then the asymptotic distribution of the system process $\mathcal{Q} = (Q_t : t \geq 0)$ for the GI/PH/1 queue is*

given by

$$y_{nj} = \begin{cases} \lambda \cdot \sum_{k=n-1}^{\infty} \mathbf{x}_k \, \psi(k+1-n) \, e_j, & n \geq 1 \\ \lambda \cdot \sum_{k=0}^{\infty} \mathbf{x}_k \sum_{h=k+1}^{\infty} \psi(h) \, \mathbf{1}, & n = 0 \end{cases}$$

with the following notations. H denotes the inter–arrival time distribution, and $H^c(t) := 1 - H(t)$ its complement. $\lambda := 1/\mathbb{E}(H)$ denotes the asymptotic arrival rate. The vector $\mathbf{x} = \mathbf{x}\tilde{P}$ contains the stationary distribution of the embedded Markov chain $\mathcal{X}$ at arrival instants, and the matrices $P_k(t)$ are given as in corollary 10.6 with $\Delta = (T, \eta\alpha, 0, 0, \ldots)$. Finally we denoted

$$\psi(k) := \int_0^{\infty} H^c(t) P_k(t) \, dt$$

for all $k \in \mathbb{N}_0$.

In the remainder of this section we shall find simpler expressions for $\mathbf{y}$. Because of theorem 11.1, we first obtain

$$\mathbf{y}_0 = \lambda \cdot \sum_{k=0}^{\infty} \mathbf{x}_k \sum_{h=k+1}^{\infty} \psi(h) \, \mathbf{1} = \lambda \cdot \mathbf{x}_0 \sum_{h=1}^{\infty} \sum_{k=0}^{h-1} R^k \psi(h) \, \mathbf{1}$$

$$= \lambda \cdot \mathbf{x}_0 (I - R)^{-1} \sum_{h=1}^{\infty} (I - R^h) \psi(h) \, \mathbf{1}$$

$$= \lambda \cdot \mathbf{x}_0 (I - R)^{-1} \left(\int_0^{\infty} H^c(t)(1 - P_0(t)\mathbf{1}) \, dt - \sum_{h=1}^{\infty} R^h \psi(h) \, \mathbf{1} \right)$$

$$= \lambda \cdot \mathbf{x}_0 (I - R)^{-1} \mathbf{1} \cdot \mathbb{E}(H) - \lambda \cdot \mathbf{x}_0 (I - R)^{-1} \sum_{h=0}^{\infty} R^h \psi(h) \, \mathbf{1}$$

Corollary 11.5 yields $\mathbf{x}_0(I - R)^{-1}\mathbf{1} = 1$. Denoting $\Psi[R] := \sum_{h=0}^{\infty} R^h \psi(h)$ and using $\lambda \cdot \mathbb{E}(H) = 1$, we can write

$$\mathbf{y}_0 = 1 - \lambda \cdot \mathbf{x}_0 (I - R)^{-1} \Psi[R] \mathbf{1} \tag{11.7}$$

For $n \in \mathbb{N}$ the same arguments yield

$$\mathbf{y}_n = \lambda \cdot \sum_{k=n-1}^{\infty} \mathbf{x}_k \, \psi(k+1-n) = \lambda \cdot \mathbf{x}_0 R^{n-1} \Psi[R] \tag{11.8}$$

Lemma 11.12 *The matrix $\Psi[R]$ is given by*

$$\Psi[R] = (R - R \, \mathbf{1}\alpha - I) \, T^{-1}$$

Proof: Differentiating equation (10.1) for the $PH(\alpha, T)$ renewal process yields

$$P_0'(t) = P_0(t)\, T \quad \text{and} \quad P_n'(t) = P_n(t)\, T + P_{n-1}(t)\, \eta\alpha \qquad (11.9)$$

for all $n \in \mathbb{N}$. From theorem 11.3 we know the relation $R = \sum_{n=0}^{\infty} R^n A_n$. For every $n \in \mathbb{N}_0$ we obtain by partial integration

$$
\begin{aligned}
A_n &= \int_0^\infty P_n(t)\, dH(t) = - \int_0^\infty P_n(t)\, dH^c(t) \\
&= - P_n(t) H^c(t)\big|_{t=0}^\infty + \int_0^\infty P_n'(t) H^c(t)\, dt \\
&= \delta_{n0} I + \psi(n)\, T + (1 - \delta_{n0}) \cdot \psi(n-1)\, \eta\alpha
\end{aligned}
$$

where δ_{n0} denotes the Kronecker function. Multiplying by R^n and summing up over all $n \in \mathbb{N}_0$ yields

$$R = I + \Psi[R]\, T + R\, \Psi[R]\, \eta\alpha \qquad (11.10)$$

Multiplying by $\mathbf{1}$ we obtain

$$(I - R)\mathbf{1} = (I - R)\, \Psi[R]\eta$$

By theorem 11.4, the matrix $(I - R)$ is invertible in case of positive recurrence. Hence $\Psi[R]\eta = \mathbf{1}$, which yields after substitution in (11.10)

$$\Psi[R]\, T = R - R\, \mathbf{1}\alpha - I$$

The statement follows now by invertibility of T.
□

Theorem 11.13 *The asymptotic distribution* $\mathbf{y}$ *of* Q *is given by*

$$\mathbf{y}_0 = 1 - \rho \quad \text{and} \quad \mathbf{y}_n = \lambda \cdot \mathbf{x}_0 (R^n\, (I - \mathbf{1}\alpha) - R^{n-1}) T^{-1}$$

for all $n \in \mathbb{N}$.

Proof: Substituting the expression of lemma 11.12 in equation (11.7) yields

$$
\begin{aligned}
\mathbf{y}_0 &= 1 - \lambda \cdot \mathbf{x}_0 \left(-(I - R)^{-1} R\, \mathbf{1}\alpha - I \right) T^{-1}\mathbf{1} \\
&= 1 - \lambda \cdot (\alpha(I - R)^{-1}\mathbf{1})^{-1} \left(\alpha(I - R)^{-1} R\, \mathbf{1} + 1 \right) \cdot (-\alpha T^{-1}\mathbf{1})
\end{aligned}
$$

because of corollary 11.5. Writing $\mathbf{1} = \alpha(I - R)^{-1}(I - R)\mathbf{1}$ we obtain

$$\mathbf{y}_0 = 1 - \lambda \cdot (-\alpha T^{-1}\mathbf{1}) = 1 - \rho$$

Regarding $\mathbf{y}_n$ with $n \in \mathbb{N}$, the statement follows immediately after substitution of the same expression for $\Psi[R]$ into formula (11.8).
□

Notes

A classical treatment of the GI/PH/1 queue is given in Neuts [65], who presents a proof for the necessity of the stability condition, too. Different proofs for the ergodicity conditions can be found in Asmussen [5] as well as in Meyn and Tweedie [59].

Tweedie [83] has shown how to generalize the matrix–geometric solution for GI/M/1 - type matrices towards operator–geometric solutions for GI/M/1 - type matrices with a general phase space.

Exercise 11.1 Verify that for the PH service time distribution being an exponential one, all results coincide with the results obtained for the GI/M/1 queue.

Exercise 11.2 For corollary 11.5, show that $B[R] = 1\alpha$.

Exercise 11.3 Show that the mean number of arrivals during a busy period is given by $\alpha(I - R)^{-1}1$.

Exercise 11.4 Show that the stationary mean number of users in the system prior to arrivals is given by

$$\bar{N}_a = \mathbf{x}_0(I - R)^{-2}1 - 1$$

Exercise 11.5 Verify equations (11.9).

Exercise 11.6 Show that the asymptotic mean number of users in the system is given by

$$\bar{N} = \rho\bar{N}_a - \lambda\mathbf{x}_0(I - R)^{-1}T^{-1}1$$

with $\bar{N}_a$ as defined in exercise 11.4.

Chapter 12

THE BMAP/G/1 QUEUE

Let $(\mathcal{N}, \mathcal{J})$ denote a BMAP with characterizing matrices $\Delta = (D_n : n \in \mathbb{N}_0)$, each matrix D_n being of dimension $m \in \mathbb{N}$. This shall model the arrival stream into the queue. The distribution function of the service time shall be denoted by H and satisfy $0 < \mathbb{E}(H) < \infty$. The service discipline is FCFS. Let $\mathcal{Q} = (Q_t : t \geq 0)$ denote the system process comprising the phase of the arrival process. Thus, $Q_t = (n, i)$ means that there are n users in the system at time t and the arrival process has phase $J_t = i$. The state space of $\mathcal{Q}$ is $E = \mathbb{N}_0 \times \{1, \ldots, m\}$.

As the analysis of the GI/PH/1 queue was similar to that of the GI/M/1 queue, we will find many similarities between the BMAP/G/1 and the M/G/1 queueing systems. To begin with, we will first construct an embedded Markov chain at the times of service completions. Define $T_0 := 0$ and T_n as the time of the nth service completion. Write $\mathcal{T} = (T_n : n \in \mathbb{N}_0)$. Let $X_n := Q_{T_n}$ for all $n \in \mathbb{N}_0$ and assume that at time zero there are no users in the system. The T_n are stopping times for $\mathcal{Q}$ and by definition X_n is a deterministic function of Q_{T_n}. The assumption $0 < \mathbb{E}(H)$ implies $T_n \to \infty$ for $n \to \infty$.

At the time instances immediately after service completions we know that the current service (if there is one) has just begun, and we need only to remember the current system state in order to predict the system state immediately after the next service completion. This implies that $\mathcal{X} = (X_n : n \in \mathbb{N}_0)$ is a Markov chain. The same property of the queue yields condition (7.5). Hence $\mathcal{Q}$ is a semi–regenerative process with embedded Markov renewal chain $(\mathcal{X}, \mathcal{T})$. Note that this time (as opposed to the embedded chain for the M/G/1 queue) the Markov chain is two–dimensional with state space $E = \mathbb{N}_0 \times \{1, \ldots, m\}$.

1.   The Embedded Markov Chain

Define the matrices

$$A_n = \int_0^\infty P_n(t)\, dH(t) \tag{12.1}$$

and

$$B_n = \sum_{k=1}^{n+1} \int_0^\infty e^{D_0 \cdot u} D_k\, du \int_0^\infty P_{n+1-k}(t)\, dH(t) \tag{12.2}$$

for all $n \in \mathbb{N}_0$, with $P_n(t)$ denoting the transition probability matrices that the BMAP counts n arrivals in time t (see corollary 10.6). The matrix A_n contains the probabilities that within a service time n users have arrived. Hence we can describe some of the transition probabilities for the Markov chain $\mathcal{X}$ by

$$\mathbb{P}(X_1 = (l+n, j) | X_0 = (l, i)) = A_{n+1;ij}$$

independently of $l \geq 1$ and for all $n \geq -1$ and $i, j \leq m$. The matrix B_n contains the probabilities that first a batch of $1 \leq k \leq n+1$ users arrives and then $n+1-k$ additional users arrive within a service time. This situation occurs whenever a service completion leaves the queueing system empty. Therefore we can write

$$\mathbb{P}(X_1 = (n, j) | X_0 = (0, i)) = B_{n;ij}$$

for all $n \geq 0$ and $i, j \leq m$. In summary we obtain for the transition probability matrix of X the block structure

$$P = \begin{pmatrix} B_0 & B_1 & B_2 & \cdots \\ A_0 & A_1 & A_2 & \cdots \\ 0 & A_0 & A_1 & \cdots \\ 0 & 0 & A_0 & \ddots \\ \vdots & \ddots & \ddots & \ddots \end{pmatrix}$$

A matrix of this structure is said to be of **M/G/1 type**, which underlines the similarity to the embedded Markov chain of the M/G/1 queue. Again we will call the first dimension n of a state (n, i) the **level**, and the second dimension i the **phase**. With respect to the levels, the Markov chain X is called **skip–free to the left**, since in one transition the level can be reduced only by one.

Simplifying the expression (12.2) and using definition (12.1) yields the relation

$$B_n = -D_0^{-1} \sum_{k=0}^{n} D_{k+1} A_{n-k} \tag{12.3}$$

between the sequences (A_n) and (B_n), which reduces the computation of the matrices B_n to a prior computation of the sequence (A_n).

2. The Matrix G

Define τ_n as the number of steps until the chain $\mathcal{X}$ reaches level n for the first time. Further define

$$G_k(i,j) := \mathbb{P}(\tau_n = k, X_k = (n,j)|X_0 = (n+1,i))$$

for all $k \geq 1$ and $i, j \leq m$. The spatial homogeneity of P implies that this definition is independent of $n \geq 0$. Define the matrices G_k of dimension $m \times m$ by their entries $G_k(i,j)$ for $i, j \leq m$, and $G := \sum_{k=1}^{\infty} G_k$. Thus the entry $G(i,j)$ denotes the probability that under the condition that we start in a level $n+1 \geq 1$ and in phase i, we reach the level n for the first time in state j.

Theorem 12.1 *If the Markov chain $\mathcal{X}$ is recurrent, then G is stochastic.*

Proof: If G is not stochastic, then there is a phase i such that

$$\sum_{k=1}^{\infty} \sum_{j=1}^{m} \mathbb{P}(\tau_n = k, X_k = (n,j)|X_0 = (n+1,i)) < 1$$

Because the transition matrix P is skip–free to the left, this means that the function f_{ij} as defined by (2.5) satisfies $f_{(n+1,i),(n,j)} < 1$ for all $n \in \mathbb{N}_0$ and $1 \leq j \leq m$.

By definition of the matrices A_n and B_n, the Markov chain $\mathcal{X}$ is irreducible. Then $\mathcal{X}$ is recurrent by definition if $f_{x,x} = 1$ for some state $x \in E$. According to equation (2.6), $f_{x,x} = 1$ can only hold for a state $x \in E$ if $f_{x,y} = 1$ holds for all states $x, y \in E$. Hence it follows that $\mathcal{X}$ is transient if G is not stochastic. $\square$

Theorem 12.2 *The matrix G satisfies the fixed point equation*

$$G = \sum_{n=0}^{\infty} A_n G^n$$

Proof: First we introduce the auxiliary matrices $G_k^{[r]}$ with entries

$$G_k^{[r]}(i,j) = \mathbb{P}(\tau_n = k, X_k = (n,j)|X_0 = (n+r,i)) \tag{12.4}$$

for all $k, r \geq 1$ and $i, j \leq m$. Again, the spatial homogeneity of P implies that this definition is independent of $n \geq 0$. By definition $G_k^{[1]} = G_k$ for all $k \geq 1$. Because P is skip–free to the left, we have $G_k^{[r]} = 0$ for $k < r$. Summing up over the number l of steps until the chain $\mathcal{X}$ reaches the next lower level for the first time, we obtain

$$G_k^{[r]} = \sum_{l=1}^{k-1} G_l G_{k-l}^{[r-1]} \tag{12.5}$$

for all $k, r \geq 1$. Further define $G^{[r]} = \sum_{k=1}^{\infty} G_k^{[r]}$ for all $r \geq 1$. Now we obtain

$$G^{[r]} = \sum_{k=1}^{\infty} \sum_{l=1}^{k-1} G_l G_{k-l}^{[r-1]} = \sum_{l=1}^{\infty} G_l \sum_{k=l+1}^{\infty} G_{k-l}^{[r-1]} = GG^{[r-1]}$$

which implies $G^{[r]} = G^r$ because of $G^{[1]} = G$. Summing up over the level reached by $\mathcal{X}$ after the first step we finally obtain

$$G = A_0 + \sum_{n=1}^{\infty} A_n G^{[n]} = \sum_{n=0}^{\infty} A_n G^n$$

which was to be proven.
$\square$

Theorem 12.3 *The matrix G can be computed by the recursion*

$$G_1 = A_0 \qquad and \qquad G_k = \sum_{n=1}^{k-1} A_n \mathcal{G}_{k-1}^{*n}$$

with $\mathcal{G} = (0, G_1, G_2, \dots)$ and the convolutional powers of $\mathcal{G}$ as defined on page 190.

Proof: The first equality follows from the definition of G_1. For the second one, a first passage argument yields

$$G_k = \sum_{n=1}^{k-1} A_n G_{k-1}^{[n]}$$

with $G_k^{[n]}$ as defined in (12.4). Thus it remains to show that $G_k^{[n]} = \mathcal{G}_k^{*n}$ for all $k \geq n \in \mathbb{N}$. This relation holds by definition for $n = 1$ and all $k \geq 1$. For $n + 1$ the induction step follows from equation (12.5).
$\square$

3.   Stationary Distribution at Service Completions

Using the matrix G from the previous section, we are now ready to derive a recursive scheme for the stationary probability vector $\mathbf{x} = \mathbf{x}P$ of P. We write

$\mathbf{x} = (\mathbf{x}_n : n \in \mathbb{N}_0)$, with $\mathbf{x}_n = (x_{ni} : 1 \leq i \leq m)$ containing the stationary probabilities for level n.

Theorem 12.4 *If $\mathcal{X}$ is positive recurrent, then the vectors $\mathbf{x}_n$ satisfy the recursion*

$$\mathbf{x}_n = \left(\mathbf{x}_0 \bar{B}_n + \sum_{k=1}^{n-1} \mathbf{x}_k \bar{A}_{n+1-k} \right) (I - \bar{A}_1)^{-1}$$

for all $n \geq 1$, with the definitions

$$\bar{B}_n := \sum_{k=n}^{\infty} B_k G^{k-n} \quad and \quad \bar{A}_n := \sum_{k=n}^{\infty} A_k G^{k-n}$$

for $n \geq 0$.

Proof: For $n \geq 1$ we consider the Markov chain $\mathcal{X}^F$ embedded in $\mathcal{X}$ at times of visits to the set $F = \{(k, i) : 0 \leq k \leq n, i \leq m\}$ of states, which is equal to levels zero through n. By the definition of $G^{[r]}$ and due to the relation $G^{[r]} = G^r$ (see the proof of theorem 12.2), the transition probability matrix P_n of $\mathcal{X}^F$ can be written as

$$P_n = \begin{pmatrix} B_0 & B_1 & B_2 & \cdots & B_{n-1} & \bar{B}_n \\ A_0 & A_1 & A_2 & \cdots & A_{n-1} & \bar{A}_n \\ 0 & A_0 & A_1 & \cdots & A_{n-2} & \bar{A}_{n-1} \\ \vdots & \ddots & \ddots & \ddots & \vdots & \vdots \\ 0 & \cdots & 0 & A_0 & A_1 & \bar{A}_2 \\ 0 & \cdots & 0 & 0 & A_0 & \bar{A}_1 \end{pmatrix}$$

By theorem 2.29 we know that the vectors $\mathbf{x}_0, \ldots, \mathbf{x}_n$ are proportional to the vectors $\mathbf{y}_0, \ldots, \mathbf{y}_n$ with $\mathbf{y} = (\mathbf{y}_0, \ldots, \mathbf{y}_n)$ satisfying $\mathbf{y} = \mathbf{y} P_n$. Hence we obtain in particular for the nth column

$$\mathbf{x}_n = \mathbf{x}_0 \bar{B}_n + \sum_{k=1}^{n} \mathbf{x}_k \bar{A}_{n+1-k} \tag{12.6}$$

According to theorem 12.1 we know that G is stochastic, and thus

$$\bar{A}_1 \mathbf{1} = \sum_{k=1}^{\infty} A_k G^{k-1} \mathbf{1} = \sum_{k=1}^{\infty} A_k \mathbf{1} = \mathbf{1} - A_0 \mathbf{1}$$

Due to corollary 10.7 and definition (12.1), all row sums of the matrix A_0 are strictly positive. This implies that all row sums of $\bar{A}_1$ are strictly less than one, and thus $I - \bar{A}_1$ is invertible by Gershgorin's circle theorem (see corollary 15.11). Therefore, we can transform the relation (12.6) to

$$\mathbf{x}_n = \left(\mathbf{x}_0 \bar{B}_n + \sum_{k=1}^{n-1} \mathbf{x}_k \bar{A}_{n+1-k}\right)(I - \bar{A}_1)^{-1}$$

which is the statement.
□

With the above recursion it remains to determine the vector $\mathbf{x}_0$ in order to obtain the stationary distribution at service completions. A simple expression for this will be obtained as a by–product of the next section.

4.   Asymptotic Distribution of the System Process

By means of the stationary distribution $\mathbf{x}$ of the embedded Markov chain $\mathcal{X}$ at service completions we can determine the asymptotic distribution $\mathbf{y}$ of the queue's system process via theorem 7.15. In order to apply this, we need to obtain the product $\mathbf{xm}$, with $\mathbf{m}$ denoting the column vector with entries

$$m_{n,i} = \mathbb{E}(T_1 | X_0 = (n, i))$$

Theorem 12.5 *The asymptotic mean time between two service completions is given by*
$$\mathbf{xm} = \mathbb{E}(H) - \mathbf{x}_0 D_0^{-1} \mathbf{1}$$

Proof: For $n > 0$ we have $m_{n,i} = \mathbb{E}(H)$, since the service does not depend on the phase of the arrival process. In order to determine the values $m_{0,i}$, define $\tau := \min\{t \geq 0 : pr_1(Q_t) > 0\}$ as the time until the first arrival. Since the arrival process is independent of the rest of the queue, equation (10.2) can be applied to yield $\mathbb{E}(\tau | X_0 = (0, i)) = -e_i D_0^{-1} \mathbf{1}$, with e_i denoting the ith row base vector. Hence we obtain

$$m_{0,i} = \mathbb{E}(\tau | X_0 = (0, i)) + \mathbb{E}(H) = \mathbb{E}(H) - e_i D_0^{-1} \mathbf{1}$$

Now the product $\mathbf{xm}$ is given by

$$\mathbf{xm} = \sum_{i=1}^{m}\left(\sum_{n=1}^{\infty} x_{ni}\mathbb{E}(H) + x_{0i}(\mathbb{E}(H) - e_i D_0^{-1}\mathbf{1})\right) = \mathbb{E}(H) - \mathbf{x}_0 D_0^{-1}\mathbf{1}$$

which was to be proven.
□

In a stationary regime, we expect that the intensities of the flow into and out of a queueing system equal each other. The asymptotic mean arrival rate is completely determined by the parameters of the BMAP and was derived in theorem 10.10 as $\lambda = \pi \sum_{k=1}^{\infty} k D_k \mathbf{1}$, with π denoting the stationary phase distribution. The latter is determined by $\pi D = 0$, with $D = \sum_{k=0}^{\infty} D_k$ denoting the generator of the phase process. The intensity of the flow out of the system can be measured in terms of the mean time between two service completions, since every user leaves the system immediately after its service is finished. Thus we would expect that in under a stationary regime $\lambda = 1/\mathbf{xm}$ holds. This shall be proven next in

Theorem 12.6 *The asymptotic mean time between service completions equals the inverse of the asymptotic mean arrival rate:*

$$\mathbb{E}(H) - \mathbf{x}_0 D_0^{-1} \mathbf{1} = \lambda^{-1}$$

Proof: Since the arrival process is independent from the rest of the queue, $\lambda \cdot (\mathbb{E}(H) - \mathbf{x}_0 D_0^{-1} \mathbf{1}) = \lambda \cdot \mathbf{xm}$ is the asymptotic mean number of arrivals between two service completions, according to theorem 12.5. We need to show $\lambda \cdot \mathbf{xm} = 1$. Using the stationary distribution $\mathbf{x}$ of the embedded Markov chain X at service completion times, we recognize the following representations: The number

$$M_1 = \sum_{n=1}^{\infty} \mathbf{x}_n \sum_{k=1}^{\infty} k A_k \mathbf{1}$$

indicates the mean number of arrivals between two service completions for the case that the prior service completion does not leave the queue empty. The number

$$M_2 = -\mathbf{x}_0 D_0^{-1} \sum_{k=1}^{\infty} k D_k \mathbf{1}$$

signifies the mean batch size of the first arrival after the prior service completion leaves the system empty. Finally,

$$M_3 = -\mathbf{x}_0 D_0^{-1} \sum_{k=1}^{\infty} D_k \sum_{k=1}^{\infty} k A_k \mathbf{1}$$

represents the mean number of arrivals during the following service time. Thus we can write $M_1 + M_2 + M_3 = \lambda \cdot \mathbf{xm}$ and it suffices to show that this sum equals one. To this aim we take a look at the defining equation $\mathbf{x} = \mathbf{x}P$. Analogously to the argument before equation (8.16) for the M/G/1 queue, this can be written as

$$\mathbf{x}_1 A_0 = \mathbf{x}_0(I - B_0)$$
$$\mathbf{x}_2 A_0 = \mathbf{x}_0(I - B_0 - B_1) + \mathbf{x}_1(I - A_0 - A_1)$$
$$\mathbf{x}_3 A_0 = \mathbf{x}_0(I - B_0 - B_1 - B_2) + \mathbf{x}_1(I - A_0 - A_1 - A_2)$$
$$+ \mathbf{x}_2(I - A_0 - A_1)$$
$$\vdots$$

Here the nth equation is equivalent to

$$\mathbf{x}_n A_0 = \mathbf{x}_0 \left(B - \sum_{k=0}^{n-1} B_k \right) + \sum_{l=1}^{n-1} \mathbf{x}_l \left(A - \sum_{k=0}^{n-l} A_k \right)$$

$$+ \mathbf{x}_0(I - B) + \sum_{l=1}^{n-1} \mathbf{x}_l(I - A)$$

with $A = \sum_{n=0}^{\infty} A_n$ and $B = \sum_{n=0}^{\infty} B_n$. We first multiply each equation by $\mathbf{1}$ from the right, then we add up all equations. Employing the relations $B - \sum_{k=0}^{n-1} B_k = \sum_{k=n}^{\infty} B_k$ and $A - \sum_{k=0}^{n-1} A_k = \sum_{k=n}^{\infty} A_k$, we obtain

$$\sum_{n=1}^{\infty} \mathbf{x}_n \left(A - \sum_{k=1}^{\infty} A_k \right) \mathbf{1} = \mathbf{x}_0 \sum_{n=1}^{\infty} n B_n \mathbf{1} + \sum_{n=1}^{\infty} \mathbf{x}_n \sum_{k=1}^{\infty} k A_{k+1} \mathbf{1}$$

Because of $\sum_{k=1}^{\infty} k A_{k+1} = \sum_{k=1}^{\infty} k A_k - \sum_{k=1}^{\infty} A_k$ and $A\mathbf{1} = \mathbf{1}$ this simplifies to

$$\mathbf{1} - \mathbf{x}_0 \mathbf{1} = \mathbf{x}_0 \sum_{n=1}^{\infty} n B_n \mathbf{1} + M_1 \tag{12.7}$$

Using $B_n = -D_0^{-1} \sum_{k=1}^{n+1} D_k A_{n+1-k}$ (see equation (12.3)), the first term on the right is evaluated as

$$\mathbf{x}_0 \sum_{n=1}^{\infty} n B_n \mathbf{1} = -\mathbf{x}_0 D_0^{-1} \sum_{n=1}^{\infty} n \sum_{k=1}^{n+1} D_k A_{n+1-k} \mathbf{1}$$

$$= -\mathbf{x}_0 D_0^{-1} \sum_{k=1}^{\infty} D_k \sum_{n=k-1}^{\infty} ((n+1-k) + (k-1)) A_{n+1-k} \mathbf{1}$$

$$= M_3 - \mathbf{x}_0 D_0^{-1} \sum_{k=1}^{\infty} (k-1) D_k \sum_{n=0}^{\infty} A_n \mathbf{1}$$

$$= M_3 + M_2 + \mathbf{x}_0 D_0^{-1} \sum_{k=1}^{\infty} D_k \mathbf{1}$$

$$= M_3 + M_2 - \mathbf{x}_0 \mathbf{1}$$

where the last equality holds because of $\sum_{k=1}^{\infty} D_k = (D - D_0)$ and $D1 = 0$.
This and (12.7) yield the statement.
□

Let $y_{ni} := \lim_{t \to \infty} \mathbb{P}(Q_t = (n, i))$ for $n \in \mathbb{N}_0$ and $1 \leq i \leq m$ denote the asymptotic probabilities of the system process Q. Further define the vectors $\mathbf{y}_n := (y_{n1}, \ldots, y_{nm})$ for all $n \in \mathbb{N}_0$ and the sequence $\mathbf{y} = (\mathbf{y}_n : n \in \mathbb{N}_0)$. Define the $m \times m$ matrices $K^{[kn]}(t)$ by their entries

$$K_{ij}^{[kn]}(t) := \mathbb{P}(T_1 > t, Q_t = (n, j) | X_0 = (k, i))$$

for all $t \geq 0$ and $k, n \in \mathbb{N}_0$. Then the asymptotic distribution $\mathbf{y}$ of the system process can be expressed in terms of the stationary distribution $\mathbf{x}$ of the embedded Markov chain $\mathcal{X}$ via theorem 7.15. As a first result we obtain

Theorem 12.7 *The asymptotic probability vector of an empty system is given by*

$$\mathbf{y}_0 = -\lambda \mathbf{x}_0 D_0^{-1}$$

and has total mass $\mathbf{y}_0 1 = 1 - \rho$, *with* $\rho = \lambda \cdot \mathbb{E}(H)$.

Proof: For $n = 0$ the matrices $K^{[kn]}(t)$ are zero if $k > 0$, since during the time between service completions the number of users in the system cannot decrease. The remaining matrices $K^{[00]}(t)$ are given by

$$K^{[00]}(t) = P_0(t) = e^{D_0 \cdot t}$$

for all $t \geq 0$, according to corollary 10.7. Theorem 7.15 now yields

$$\mathbf{y}_0 = \frac{1}{\mathbf{x}m} \mathbf{x}_0 \int_0^{\infty} e^{D_0 \cdot t} \, dt = \lambda \cdot \mathbf{x}_0 (-D_0^{-1})$$

which is the first statement. The second one follows from this representation of $\mathbf{y}_0$ and theorem 12.6.
□

Now we will derive a simple expression for the vector $\mathbf{x}_0$. To this aim we take a look at the Markov chain $\mathcal{X}^0$ embedded in $\mathcal{X}$ at visits to the level zero. The state space of this chain is $\{(0, i) : 1 \leq i \leq m\}$, which is isomorphic to the phase space $\{1, \ldots, m\}$ of the BMAP. From theorem 2.29 we know that $\mathbf{x}_0$ is proportional to, i.e. a scalar multiplicative of, the stationary probability vector of $\mathcal{X}^0$. First we determine the transition probability matrix K for $\mathcal{X}^0$, which is of dimension $m \times m$. From the definition of $G^{[r]}$ and the relation $G^{[r]} = G^r$

we see by a first passage argument and then by equation (12.3) that

$$K = \sum_{n=0}^{\infty} B_n G^n = -D_0^{-1} \sum_{n=0}^{\infty} \sum_{k=0}^{n} D_{k+1} A_{n-k} G^n$$

$$= -D_0^{-1} \sum_{k=0}^{\infty} D_{k+1} \sum_{n=k}^{\infty} A_{n-k} G^{n-k} G^k = -D_0^{-1} \sum_{k=0}^{\infty} D_{k+1} G^{k+1}$$

$$= -D_0^{-1} \left(\sum_{k=0}^{\infty} D_k G^k - D_0 \right) = I - D_0^{-1} D[G]$$

with $D[G] := \sum_{k=0}^{\infty} D_k G^k$. In order to find an expression for the vector $\kappa = \kappa K$, we first need the following representation of the matrix G:

Lemma 12.8 *The matrix G can be expressed by*

$$G = \int_0^{\infty} e^{D[G] \cdot t} dH(t)$$

In particular, the invariant probability vector $\mathbf{g} = \mathbf{g}G$ satisfies $\mathbf{g}D[G] = 0$.

Proof: The matrix G contains the probabilities of phase transitions between a service completion that does not leave the system empty and the first consecutive service completion which leaves the system with one user less than at the beginning. During the time between these two service completions the queue is never empty, which means that this time interval is a finite sum of (randomly many) service times.

Phase transitions depend on the arrival process only, since this is independent of the rest of the queue. Thus it does not matter for G which is the service discipline, as long as the time between the above mentioned service completions remains the same.

The stated expression for G results if we regard the phase process under the following service discipline. Whenever a new user arrives, it is immediately admitted to the server. The current service is interrupted and the user in service goes to the head of the queue. As soon as a service is completed, the service of the user at the head of the queue is resumed, i.e. none of the work is lost.

Thus the time that a user spends in the server still equals exactly its service time. The server is not idle between the above mentioned service completions and finally, since the arrival process is independent from the service, the number of arriving users does not change under the new service discipline.

Under the new service discipline, the user that is in service at the beginning of the time interval concerning G will also be in service when this time interval ends, since all users arriving later will be served earlier. If there are no arrivals during the service time of this user, then phase transitions are governed by the rate matrix D_0. If there is a first (batch) arrival, occuring with rate matrix D_n, then the phase upon reentering the same level again (when the first user resumes its service) will change according to the rate matrix $D_n G^{[n]} = D_n G^n$. Thus the generator for the phase process, if we regard only the lowest level of the first user in service, is given by $D[G] = \sum_{n=0}^{\infty} D_n G^n$. Since the complete time that the first user spends in the server is exactly its service time, the expression for G follows.

The second statement $\mathbf{g}D[G] = 0$ for the stationary probability vector $\mathbf{g} = \mathbf{g}G$ follows immediatlely from the obtained representation for G.
□

Remark 12.9 The service discipline that was involved in the above proof is called LCFS (last come first served) discipline with preemptive resume regulation.

Theorem 12.10 *The stationary probabilities that the chain $\mathcal{X}$ is in level zero can be expressed by*

$$\mathbf{x}_0 = -\frac{1-\rho}{\lambda} \mathbf{g} D_0$$

and the asymptotic probability vector of an empty system is given by

$$\mathbf{y}_0 = (1-\rho) \cdot \mathbf{g}$$

Proof: The expression $K = I - D_0^{-1} D[G]$ along with lemma 12.8 yields that $\kappa = c' \cdot (-\mathbf{g}D_0)$ with some constant c'. Theorem 2.29 now states that a representation $\mathbf{x}_0 = c \cdot (-\mathbf{g}D_0)$ holds with some constant c. By theorem 12.7 this implies

$$1 - \rho = \mathbf{y}_0 \mathbf{1} = \lambda \cdot c \cdot \mathbf{g} D_0 D_0^{-1} \mathbf{1} = \lambda \cdot c$$

and thus $c = (1-\rho)/\lambda$. This proves the first statement. The second statement now is a consequence of theorem 12.7.
□

Theorem 12.11 *The asymptotic probability vectors $\mathbf{y}_n$ for $n \geq 1$ are given by the recursion*

$$\mathbf{y}_n = \sum_{k=1}^{n} (\mathbf{y}_0 D_k + \lambda \mathbf{x}_k) \int_0^{\infty} (1 - H(t)) P_{n-k}(t) \, dt$$

Proof: An application of theorem 7.15, along with theorems 12.5 and 12.6, yields

$$\mathbf{y}_n = \lambda \sum_{k=0}^{n} \mathbf{x}_k \int_0^{\infty} K^{[kn]}(t) \, dt \tag{12.8}$$

as between service completions the number of users in the system can only increase. For $k = 0$ and $n \geq 1$ we obtain

$$K^{[0n]}(t) = \int_0^t e^{D_0 \cdot u} \sum_{l=1}^{n} D_l \cdot (1 - H(t-u)) \cdot P_{n-l}(t-u) \, du$$

while for $k, n > 0$ we have

$$K^{[kn]}(t) = (1 - H(t)) \cdot P_{n-k}(t)$$

In both expressions the independence between arrival process and current service is used. Employing them in (12.8) yields

$$\mathbf{y}_n = \lambda \cdot \mathbf{x}_0 \sum_{l=1}^{n} \int_{t=0}^{\infty} \int_{u=0}^{t} e^{D_0 \cdot u} D_l \cdot (1 - H(t-u)) \cdot P_{n-l}(t-u) \, du \, dt$$

$$+ \lambda \cdot \sum_{k=1}^{n} \mathbf{x}_k \int_0^{\infty} (1 - H(t)) \cdot P_{n-k}(t) \, dt$$

The integral in the first line equals

$$\int_{u=0}^{\infty} e^{D_0 \cdot u} D_l \int_{t=u}^{\infty} (1 - H(t-u)) P_{n-l}(t-u) \, dt \, du$$

$$= \int_{u=0}^{\infty} e^{D_0 \cdot u} \, du \, D_l \int_{t=0}^{\infty} (1 - H(t)) P_{n-l}(t) \, dt$$

$$= -D_0^{-1} D_l \int_{t=0}^{\infty} (1 - H(t)) P_{n-l}(t) \, dt$$

If we plug this back into the above expression for $\mathbf{y}_n$ and use $\mathbf{y}_0 = -\lambda \mathbf{x}_0 D_0^{-1}$, we obtain the statement.

$\square$

5.   Stability Conditions

As in the previous chapter on the GI/PH/1 queue we will show various stability conditions to be equivalent. Define $A = \sum_{n=0}^{\infty} A_n$. Then $A = (a_{ij})_{i,j \leq m}$ is the transition matrix for the phase component of the embedded Markov chain $\mathcal{X}$ under the condition that the queue is not empty. More exactly, we have

$$\mathbb{P}(pr_2(X_1) = j | X_0 = (n, i)) = a_{ij}$$

for all $n \geq 1$. Denote the stationary probability vector of A by $\pi = \pi A$. Further denote the generator of the phase process J by $D = \sum_{n=0}^{\infty} D_n$. Completely analogously to the proof of lemma 11.7 one can show (as an exercise) that the vector π satisfies $\pi D = 0$.

Theorem 12.12 *Denote the mean service time by* $\mathbb{E}(H)$ *and the asymptotic mean arrival rate of the BMAP by* $\lambda = \pi \sum_{n=1}^{\infty} n D_n \mathbf{1}$. *Then*

$$\rho = \lambda \cdot \mathbb{E}(H) = \pi \sum_{n=1}^{\infty} n A_n \mathbf{1}$$

Proof: Simply using the definition of A_n yields

$$\pi \sum_{n=1}^{\infty} n A_n \mathbf{1} = \pi \sum_{n=1}^{\infty} n \int_0^{\infty} P_n(t) \, dH(t) \mathbf{1} = \int_0^{\infty} \mathbb{E}_\pi(N_t) \, dH(t)$$

where $\mathbb{E}_\pi(N_t)$ is the expected number of arrivals during time t if the BMAP starts with a phase distribution π. By theorem 10.10 we have $\mathbb{E}_\pi(N_t) = \lambda \cdot t$ and hence

$$\pi \sum_{n=1}^{\infty} n A_n \mathbf{1} = \lambda \cdot \int_0^{\infty} t \, dH(t) = \lambda \cdot \mathbb{E}(H)$$

which is the statement.
$\square$

Theorem 12.13 *If the stability condition* $\rho < 1$ *holds, then the embedded Markov chain* $\mathcal{X}$ *is positive recurrent.*

Proof: As in lemma 11.9 we can find a non–negative solution $\mathbf{x}$ to the equation system

$$(I - A)\mathbf{x} = \bar{\mathbf{a}} - \rho \mathbf{1} \tag{12.9}$$

with $\bar{a}_i := \sum_{j \in E} \sum_{n=1}^{\infty} n A_{n;ij}$ for $i \in E$. Define the function $f(s, j) = s + x_j$ for $s \in \mathbb{N}_0$ and $1 \leq j \leq m$. Then for $r > 0$ we obtain

$$\sum_{(s,j)} \tilde{P}_{(r,i),(s,j)} f(s, j) - f(r, i) = \sum_{j=1}^{m} \sum_{n=0}^{\infty} A_{n;ij} \cdot (r - 1 + n + x_j) - r - x_i$$

$$= r - 1 + \sum_{j=1}^{m} \sum_{n=1}^{\infty} n A_{n;ij} + \sum_{j=1}^{m} \sum_{n=0}^{\infty} A_{n;ij} x_j - r - x_i$$

$$= \sum_{j=1}^{m} \sum_{n=1}^{\infty} n A_n(i, j) + \rho - \bar{a}_i - 1 = \rho - 1 < 0$$

For the exceptional set $F = \{(0, i) : 1 \leq i \leq m\}$ we obtain

$$\sum_{(s,j)} \tilde{P}_{(0,i),(s,j)} f(s, j) = \sum_{j=1}^{m} \sum_{n=1}^{\infty} n B_{n;ij} + \sum_{j=1}^{m} \sum_{n=0}^{\infty} B_{n;ij} x_j$$

which is finite by assumption. Thus the conditions of theorem 2.33 are satisfied, which proves the statement.

□

Notes

The first complete analysis of the BMAP/G/1 queue has appeared in a paper by Ramaswami [71]. In this paper the BMAP was used with its older, more complicated notation under the name N–process. A special case of the BMAP/G/1 queue, namely the MAP/G/1 queue without batch arrivals, has been analyzed in Lucantoni et al. [56] using the current notation. The recursion scheme for the stationary probability vectors at service completion times has been introduced by Ramaswami [72]. An outline of Ramaswami's analysis using the new notations, along with some new results (namely lemma 12.8 and theorem 12.10), are presented in Lucantoni [54]. The use of a matrix convolutional calculus for the determination of the matrix G has been presented in Baum [9]. A general discussion of M/G/1 type matrices and their use in queueing theory is presented in Neuts [66], including necessary conditions for the stability of the queue. A variant of the MAP/G/1 queue with LCFS service discipline is analyzed in Breuer [20]. For a historical overview of the developments that led to the BMAP and matrix–analytical methods see Lucantoni [55].

A different proof of theorem 12.6 can be found in Ramaswami [71]. In Neuts [66] a computation of the matrix G is proposed via the fixed point equation of theorem 12.2. A more elaborate version of the proof for lemma 12.8 can be found in Lucantoni and Neuts [57], while the idea for this proof has been presented in an earlier form of notation by Machihara [58]. A more elementary proof is given in Lucantoni [54]. Another recursion scheme for the asymptotic distribution y is presented in Takine [82].

Exercise 12.1 Prove $\pi D = 0$ for the stationary distribution $\pi = \pi A$.

Exercise 12.2 Show the existence of a solution x to equation (12.9).

Exercise 12.3 Define the z–transforms $X(z) := \sum_{n=0}^{\infty} \mathbf{x}_n z^n$ of the stationary probability vector at service completions, as well as $A(z) := \sum_{n=0}^{\infty} A_n z^n$, and

$B(z) := \sum_{n=0}^{\infty} B_n z^n$ for $|z| \le 1$.

(a) Show that

$$B(z) = -z^{-1} D_0^{-1}(D(z) - D_0)A(z)$$

(b) Use the result above and $x = xP$ to show that

$$X(z)(zI - A(z)) = -x_0 D_0^{-1} D(z) A(z)$$

Exercise 12.4 Use exercise 10.2 to show that

$$\sum_{n=0}^{\infty} P_n(t) z^n = e^{D(z) \cdot t}$$

Exercise 12.5 Show that $D(z)$ is invertible for $0 \le z < 1$.

Exercise 12.6 Define $\psi(n) := \int_0^{\infty} (1 - H(t)) P_n(t) dt$ for all $n \in \mathbb{N}_0$ and $\Psi(z) := \sum_{n=0}^{\infty} \psi(n) z^n$. Show that

$$\Psi(z) = (A(z) - I) D(z)^{-1}$$

Exercise 12.7 Define the z–transform $Y(z) := \sum_{n=0}^{\infty} y_n z^n$ of the asymptotic probability vector. Show that

$$Y(z) = \begin{cases} \lambda X(z) \cdot (z - 1) \cdot D(z)^{-1}, & 0 \le z < 1 \\ \pi, & z = 1 \end{cases}$$

Hint: Start by transforming $Y(z) - y_0$ and use exercise 12.6.

Chapter 13

DISCRETE TIME APPROACHES

1. Discrete Phase–Type Distributions

Analogous to the definition of PH distributions in continuous time, we will define discrete PH distributions in terms of Markov chains with one absorbing state. Let $\mathcal{X}$ denote a Markov chain with finite state space $E = \{0, \ldots, m\}$ and a transition matrix structured as

$$P = \begin{pmatrix} 1 & 0 \\ \eta & T \end{pmatrix}$$

Denote the initial distribution of $\mathcal{X}$ by the row vector $\tilde{\alpha} = (\alpha_0, \alpha)$, where α is of dimension m. The structure of P shows that state 0 is absorbing. All other states shall be transient. Let

$$Z = \min\{n \in \mathbb{N}_0 : X_n = 0\}$$

denote the time until absorption in state 0. Define $p_n := \mathbb{P}(Z = n)$ for all $n \in \mathbb{N}_0$. The distribution $\mathbf{p} = (p_n : n \in \mathbb{N}_0)$ of Z is called a **discrete phase–type distribution**, or shortly discrete PH distribution. We also write $Z \sim PH_d(\alpha, T)$. The number m of transient states is called the **order** of $\mathbf{p}$. A transient state is called **phase**.

An immediate first observation is $\eta = 1 - T\mathbf{1}$, with $\mathbf{1}$ denoting a column vector with all entries being one. Further the definition yields

$$p_0 = \mathbb{P}(Z = 0) = \alpha_0 = 1 - \alpha\mathbf{1} \tag{13.1}$$

This explains the notation $PH_d(\alpha, T)$. Knowledge of α and T is sufficient to determine $\tilde{\alpha}$ and η and hence completely specify the distribution of Z. Therefore we call the pair (α, T) the **characterization** of a discrete PH distribution.

Theorem 13.1 *Let Z denote a random variable which has a discrete phase–type distribution with characterization (α, T). Then*

$$\mathbb{P}(Z = n) = \alpha T^{n-1} \eta \qquad and \qquad \mathbb{P}(Z \leq n) = 1 - \alpha T^n \mathbf{1}$$

for all $n \in \mathbb{N}$.

Proof: The structure of P leads to the observation

$$P^n = \begin{pmatrix} 1 & 0 \\ 1 - T^n \mathbf{1} & T^n \end{pmatrix}$$

for $n \in \mathbb{N}_0$, which can be verified by induction on n. Together with (13.1) this yields the second statement. The first one is now obtained as

$$\mathbb{P}(Z = n) = \mathbb{P}(Z \leq n) - \mathbb{P}(Z \leq n - 1) = \alpha T^{n-1} \mathbf{1} - \alpha T^n \mathbf{1}$$
$$= \alpha T^{n-1}(\mathbf{1} - T\mathbf{1})$$

which completes the proof because of $\eta = \mathbf{1} - T\mathbf{1}$.
$\square$

By corollary 2.15 we know that invertibility of $I - T$ is equivalent to the postulate that the states $1, \ldots, m$ be transient. The same arguments as in the continuous case (see theorem 9.3) serve to show that

Lemma 13.2 *A $PH_d(\alpha, T)$ distribution is non–defective if and only if the matrix $I - T$ is invertible. Then the expected number E_{ij} of visits to state j before absorption, given that the Markov chain X starts in state i, is $E_{ij} = (I - T)_{ij}^{-1}$.*

As already stated in the definition, we shall always assume that $1, \ldots, m$ are transient states, i.e. that $I - T$ is invertible. The following examples show the high versatility of the introduced class of distributions.

Example 13.3 Let $\mathbf{p} = (p_n : n \in \mathbb{N})$ denote a geometric distribution with parameter q, i.e. $p_n = (1 - q)q^{n-1}$ for all $n \in \mathbb{N}$. Then $\mathbf{p}$ has a discrete PH representation with order $m = 1$, $\alpha = 1$, and $T = q$. The exit vector is given by $\eta = 1 - q$.

Example 13.4 A generalization of the geometric distribution is the negative binomial distribution. For parameters $N \in \mathbb{N}$, the number of successes sought, and $q \in]0, 1[$, the probability of success, a distribution $\mathbf{p}$ is negative binomial if $p_n = \binom{N+n-1}{n} q^N (1 - q)^{n-N}$ for all $n \geq N$. The value p_n is the probability of observing n trials until the Nth success. For the special case $N = 1$ we obtain the geometric distribution.

The distribution p has a discrete PH representation with order $m = N$, initial phase distribution $\alpha = e_1 = (1, 0, \ldots, 0)$, and T given by the entries

$$
T_{ij} = \begin{cases} 1 - q, & i = j \\ q, & j = i + 1 \leq N \\ 0, & \text{else} \end{cases}
$$

The exit vector is $\eta = (0, \ldots, 0, q)^T$.

Example 13.5 Any discrete distribution p with finite support, i.e. $p_n = 0$ for $n > m$ with some $m \in \mathbb{N}$, has a discrete PH representation. We write $\mathbf{p} = (p_0, \ldots, p_m)$. Then there are two possibilities for such a representation.

One is called the **remaining time representation**. Here we set $\tilde{\alpha} = \mathbf{p}$, $T_{i,i-1} = 1$ for $1 \leq i \leq m$, and $T_{ij} = 0$ otherwise. This implies an exit vector $\eta = (1, 0, \ldots, 0)$.

The other is called the **elapsed time representation**. For this we set $\alpha_0 = p_0$, $\alpha = (1 - p_0, 0, \ldots, 0)$ and

$$
T_{ij} = \begin{cases} 1 - p_i / (1 - \sum_{k=0}^{i-1} p_k), & j = i + 1 \leq m \\ 0, & \text{else} \end{cases}
$$

This time we have an exit vector $\eta = (\eta_1, \ldots, \eta_{m-1}, 1)$ with entries determined by $\eta_i = p_i / (1 - \sum_{k=0}^{i-1} p_k)$.

The z–transform of a discrete phase–type distribution p is given by

$$
\mathbf{p}^*(z) = \sum_{n=0}^{\infty} p_n z^n = \alpha_0 + \sum_{n=1}^{\infty} \alpha T^{n-1} \eta z^n = \alpha_0 + z \cdot \alpha \sum_{n=1}^{\infty} (zT)^{n-1} \eta
$$
$$
= \alpha_0 + z\alpha (I - zT)^{-1} \eta \tag{13.2}
$$

for $|z| \leq 1$. This expression yields the factorial moments for a random variable $Z \sim PH_d(\alpha, T)$, namely

$$
\mathbb{E}(Z \cdot (Z - 1) \cdot \ldots \cdot (Z - k + 1)) = k! \alpha (I - T)^{-k} T^{k-1} \mathbf{1} \tag{13.3}
$$

for all $k \in \mathbb{N}$. This formula is obtained by differentiating (13.2) k times with respect to z and evaluating at $z = 1$. In particular, the mean time to absorption is given by

$$
\mathbb{E}(Z) = \alpha (I - T)^{-1} \mathbf{1} \tag{13.4}
$$

Another expression for this will be derived in corollary 13.6.

2.   BMAPs in Discrete Time

Like the discrete time version of phase–type distributions, we can define batch Markovian arrival processes in discrete time, too. Let $\mathcal{Y} = (\mathcal{N}, \mathcal{J})$ denote a Markov chain with state space

$$E = \mathbb{N}_0 \times \{1, \ldots, m\}$$

where $m \in \mathbb{N}$ is some finite number. For a state (n, i) we call the first dimension n the **level** and the second dimension i the **phase**. Let 0 denote the matrix with all entries being zero. If the transition matrix of $\mathcal{Y}$ has a block structure

$$P = \begin{pmatrix} D_0 & D_1 & D_2 & D_3 & \cdots \\ 0 & D_0 & D_1 & D_2 & \ddots \\ 0 & 0 & D_0 & D_1 & \ddots \\ \vdots & \ddots & \ddots & \ddots & \ddots \end{pmatrix}$$

and $D := \sum_{n=0}^{\infty} D_n$ is irreducible, then $\mathcal{Y}$ is called a **discrete batch Markovian arrival process** or shortly **discrete BMAP**. The Markov chain determined by the transition matrix D is called the **phase process** of $\mathcal{Y}$.

Like the continuous time analogue, we want to use discrete BMAPs as a model for arrival streams. Thus we always assume that D_0 is strictly substochastic, i.e. there is an index $n \in \mathbb{N}$ with $D_n \neq 0$. Then $\mathcal{Y}$ is clearly transient and does not have a stationary distribution.

The Toeplitz structure of P implies that the transition probabilities

$$\mathbb{P}(Y_1 = (n + k, j) | Y_0 = (n, i)) = D_k(i, j) = \mathbb{P}(Y_1 = (k, j) | Y_0 = (0, i))$$

are homogeneous in the first dimension of the state space. Hence the n–step transition probabilities are determined by the values

$$P_{k;i,j}(n) := \mathbb{P}(Y_n = (k, j) | Y_0 = (0, i))$$

Define the $m \times m$ matrices $P_k(n) := (P_{k;i,j}(n))_{i,j \leq m}$ and the sequences $P(n) := (P_k(n) : k \in \mathbb{N}_0)$ of matrices. By definition

$$P(0) = (I, 0, 0, \ldots)$$

with I denoting the $m \times m$ identity matrix. Define convolutions of matrix sequences as in section 4. Further write $\Delta = (D_n : n \in \mathbb{N}_0)$. Then clearly $P(1) = \Delta$, and we can show that

$$P(n) = \Delta^{*n}$$

for all $n \in \mathbb{N}$ by the induction step

$$P_k(n+1) = \sum_{i=0}^{k} P_i(n) D_{k-i} = \sum_{i=0}^{k} \Delta_i^{*n} D_{k-i} = \Delta_k^{*(n+1)}$$

which holds for all $k \in \mathbb{N}_0$. The z–transform of $P(n)$ is given by

$$P_n^*(z) = \sum_{k=0}^{\infty} P_k(n) z^k = \sum_{k=0}^{\infty} \Delta_k^{*n} z^k = \left(\sum_{k=0}^{\infty} D_k z^k \right)^n$$

for all $n \in \mathbb{N}$. Hence the expectation matrix of the number of arrivals within n time slots is

$$\mathbb{E}(P(n)) = \frac{\partial}{\partial z} P_n^*(z) \Big|_{z=1} = n \cdot D \cdot \sum_{k=1}^{\infty} k D_k$$

Let $\pi = \pi D$ denote the stationary distribution of the phase process. Then the expected number of arrivals within n time slots given that $\mathcal{Y}$ starts with phase distribution π is obtained as

$$\mathbb{E}_\pi(N_n) = n \cdot \pi \sum_{k=1}^{\infty} k D_k \mathbf{1} \tag{13.5}$$

for all $n \in \mathbb{N}$.

Consider now a discrete phase–type distribution with characterization (α, T). As usual, define the exit vector by $\eta := 1 - T\mathbf{1}$. A special class of discrete BMAPs arises if we set $D_0 := T$, $D_1 := \eta\alpha$ and $D_n := 0$ for $n \geq 2$. This is called a **discrete PH renewal process** or shortly a PH_d renewal process. For the stationary phase distribution $\pi = \pi D$ with $D = T\eta\alpha$, expression (13.5) specifies to $\mathbb{E}_\pi(N_n) = n \cdot \pi\eta$ for all $n \in \mathbb{N}$. Now the same argument as for corollary 10.12 holds. The described BMAP is a renewal process (in continuous time, denoted by $\tilde{N}$) with initial delay $X_0 \sim PH_d(\pi, T)$ and renewal intervals $X_n \sim PH_d(\alpha, T)$. The elementary renewal theorem 6.12 states that

$$\lim_{t \to \infty} \frac{\mathbb{E}(\tilde{N}_t)}{t} = \lim_{t \to \infty} \frac{\mathbb{E}(\tilde{N}_{\lfloor t \rfloor})}{\lfloor t \rfloor} \frac{\lfloor t \rfloor}{t} = \lim_{n \to \infty} \frac{\mathbb{E}_\pi(N_n)}{n} = \frac{1}{\mathbb{E}(X_1)}$$

Thus we obtain another expression for the mean of a discrete phase–type distributed random variable (cf. corollary 13.6).

Corollary 13.6 *For a $PH_d(\alpha, T)$ distributed random variable X the expectation is given by $\mathbb{E}(X) = (\pi\eta)^{-1}$, where $\pi = \pi(T + \eta\alpha)$ is the stationary phase distribution.*

3.   Blockwise Skip–Free Markov Chains

In chapters 11 and 12 we have analyzed Markov chains of a blockwise Hessenberg structure, i.e. they were blockwise skip–free in one direction. For each of them we have developed an own method of finding the stationary distribution. Both methods employed matrices of central importance for the formulation of the stationary distribution. In the former case it was an expectation matrix called R, in the latter case a stochastic matrix called G.

For the special case that a Markov chain is blockwise skip–free in both directions, we can hope to combine both approaches and thus obtain more results. This shall be pursued in this section. We further will see that this kind of Markov chains can be used as the basic tool for analyzing queues in discrete time.

An irreducible transition matrix structured as

$$P = \begin{pmatrix} B & C & & & \\ D & A_1 & A_0 & & \\ & A_2 & A_1 & A_0 & \\ & & A_2 & A_1 & A_0 \\ & & & \ddots & \ddots & \ddots \end{pmatrix}$$

with matrices B, C, D, and A_i having dimensions $n \times n$, $n \times m$, $m \times n$, and $m \times m$, respectively, is called **blockwise skip–free**. This matrix defines a Markov chain $\mathcal{X}$ with state space

$$E = \{(0, i) : 1 \leq i \leq n\} \cup \{(n, j) : n \in \mathbb{N}, 1 \leq j \leq m\}$$

with $n, m \in \mathbb{N}$. The first dimension of a state is called **level**, the second **phase**. The special case $n = m = 1$ is called a **skip–free Markov chain** (cf. section 2 for the continuous time analogue).

The matrix P satisfies the conditions of blockwise Hessenberg structure in both directions. Hence the approaches for analyzing the embedded Markov chains in chapters 11 and 12 apply both. The only difference to be considered are the matrices B, C, and D at the boundary.

As in section 2 we can define a rate matrix R which satisfies

$$R = A_0 + RA_1 + R^2 A_2$$

according to theorem 11.3. Theorem 11.1 tells us that if there is a stationary distribution $\mathbf{x} = (\mathbf{x}_n : n \in \mathbb{N}_0)$ for $\mathcal{X}$, then it satisfies the relation

$$\mathbf{x}_{n+1} = \mathbf{x}_n R \qquad \text{or} \qquad \mathbf{x}_n = \mathbf{x}_1 R^{n-1} \tag{13.6}$$

for all $n \in \mathbb{N}$. On the other hand we can define a matrix G as in section 2 by

$$G = A_2 + A_1 G + A_0 G^2$$

due to theorem 12.2. According to theorem 12.1 we know that G is stochastic if $\mathcal{X}$ is recurrent.

Equation (13.6) reduces the problem of finding a stationary distribution $\mathbf{x}$ for P to the determination of $\mathbf{x}_0$ and $\mathbf{x}_1$. To this aim, we consider the Markov chain $\mathcal{X}^F$ restricted to the subset

$$F = \{(0, i) : 1 \leq i \leq n\} \cup \{(1, j) : 1 \leq j \leq m\}$$

of the state space E. Because P is blockwise skip–free, states in level 0 do not communicate with states in $F^c = E \setminus F$. Hence transitions from F to F^c and back must go via level 1. Thus we arrive at a transition matrix

$$P^F = \begin{pmatrix} B & C \\ D & U \end{pmatrix}$$

for $\mathcal{X}^F$, where only the lower right–hand entry remains to be determined. The matrix U contains all probabilities to go from level 1 back to level 1 in a finite number of steps without entering level 0.

Clearly the probabilities for one step are contained in A_1, whence we obtain $U = A_1 + U'$. The respective probabilities for more than one step (which are contained in U') must consider visits to the set F^c in all but the last step. The blockwise skip–free structure of P implies that the first step must go from level 1 to level 2, for which the transition probabilities are contained in A_0. Then we need the probabilities to go from level 2 back to level 1 in a finite number of steps. By definition these are contained in the matrix G (see section 2). Hence we obtain

$$U = A_1 + A_0 G$$

and thus have determined P^F completely.

Theorem 2.29 states that positive recurrence of $\mathcal{X}$ implies positive recurrence of $\mathcal{X}^F$. Hence P^F admits a stationary distribution $\mathbf{x}^F = (\mathbf{x}_0^F, \mathbf{x}_1^F)$ as the solution of the linear equation system

$$(\mathbf{x}_0^F, \mathbf{x}_1^F) = (\mathbf{x}_0^F B + \mathbf{x}_1^F D, \mathbf{x}_0^F C + \mathbf{x}_1^F U)$$

Define $c := \mathbf{x}_0^F \mathbf{1} + \mathbf{x}_1^F (I - R)^{-1} \mathbf{1}$. Then theorem 2.29 yields that

$$\mathbf{x}_0 = c^{-1} \mathbf{x}_0^F, \quad \mathbf{x}_1 = c^{-1} \mathbf{x}_1^F, \quad \mathbf{x}_{n+1} = \mathbf{x}_1 R^n$$

for all $n \in \mathbb{N}$ is the stationary distribution of $\mathcal{X}$. In fact, we can verify

$$\sum_{n=0}^{\infty} \mathbf{x}_n \mathbf{1} = \mathbf{x}_0 \mathbf{1} + \mathbf{x}_1 \sum_{n=0}^{\infty} R^n \mathbf{1} = c^{-1}(\mathbf{x}_0^F \mathbf{1} + \mathbf{x}_1^F (I - R)^{-1} \mathbf{1}) = 1$$

and theorem 2.29 states that $(\mathbf{x}_0, \mathbf{x}_1)$ and $(\mathbf{x}_0^F, \mathbf{x}_1^F)$ differ by a constant multiple only.

Define $A := A_0 + A_1 + A_2$. Since P is irreducible and stochastic, so is A, and thus there is a stationary distribution $\pi = \pi A$. By theorems 11.10 and 11.8 we know that $\mathcal{X}$ is positive recurrent if the condition

$$\pi A_1 \mathbf{1} + 2 \cdot \pi A_2 \mathbf{1} > 1$$

holds. Using the definition of A we obtain

$$1 = \pi A \mathbf{1} = \pi A_0 \mathbf{1} + \pi A_1 \mathbf{1} + \pi A_2 \mathbf{1}$$

which yields the equivalent condition

$$\pi A_0 \mathbf{1} < \pi A_2 \mathbf{1} \tag{13.7}$$

for positive recurrence of $\mathcal{X}$.

4.   The PH/PH/1 Queue in Discrete Time

As an application we shall analyze the PH/PH/1 queue in discrete time. Inter–arrival times as well as service times are iid and have a discrete phase–type distribution, named A and B respectively. The former has characterization (α, T) of order n, the latter (β, S) of order m. We set $\alpha_0 = \beta_0 = 0$ in order to avoid batch arrivals and instantaneous services. Denote the exit vectors by $\eta = 1 - T\mathbf{1}$ and $\zeta = 1 - S\mathbf{1}$.

Note that by example 13.5, the discrete time GI/G/1 queue is a special case of the PH/PH/1 queue if inter–arrival and service time distributions have finite support. By example 13.3, the M/M/1 queue in discrete time as examined in section 6 is a special case, too.

For any time index $n \in \mathbb{N}_0$, define the random variables N_n as the number of users in the system, K_n as the phase for the inter–arrival time, and J_n as the phase for the service time. Then the system process $\mathcal{Q} = ((N_n, K_n, J_n) : n \in \mathbb{N}_0)$ is a Markov chain with state space

$$E = \{(0, k) : 1 \le k \le n\} \cup \{(n, k, j) : n \in \mathbb{N}, 1 \le k \le n, 1 \le j \le m\}$$

and transition matrix

$$P = \begin{pmatrix} B & C & & & \\ D & A_1 & A_0 & & \\ & A_2 & A_1 & A_0 & \\ & & A_2 & A_1 & A_0 \\ & & & \ddots & \ddots & \ddots \end{pmatrix}$$

where

$$B = T, \qquad C = (\eta\alpha) \otimes \beta, \qquad D = T \otimes \zeta$$
$$A_0 = (\eta\alpha) \otimes S, \quad A_1 = T \otimes S + (\eta\alpha) \otimes (\zeta\beta), \quad A_2 = T \otimes (\zeta\beta)$$

Here the composition $\otimes$ represents the Kronecker product, which is defined in the pretext of theorem 9.13. The matrices B, C, D, and A_i are of dimensions $n \times n$, $n \times nm$, $nm \times n$, and $nm \times nm$, respectively. We see that P is blockwise skip–free which allows us to use results from the preceding section.

Define $A := A_0 + A_1 + A_2$ and let $\pi = \pi A$ denote the stationary distribution for A. Using exercices 13.4 and 13.5, we obtain

$$A = (\eta\alpha) \otimes S + T \otimes S + (\eta\alpha) \otimes (\zeta\beta) + T \otimes (\zeta\beta)$$
$$= (\eta\alpha + T) \otimes S + (\eta\alpha + T) \otimes (\zeta\beta) \tag{13.8}$$
$$= (T + \eta\alpha) \otimes (S + \zeta\beta) \tag{13.9}$$

This yields

$$\pi = \alpha^* \otimes \beta^* \tag{13.10}$$

with $\alpha^* = \alpha^*(T + \eta\alpha)$ and $\beta^* = \beta^*(S + \zeta\beta)$. Condition (13.7) for positive recurrence of Q specifies to

$$(\alpha^* \otimes \beta^*)(\eta\alpha \otimes S)\mathbf{1} < (\alpha^* \otimes \beta^*)(T \otimes \zeta\beta)\mathbf{1}$$
$$\Longleftrightarrow \quad \alpha^*\eta\alpha\mathbf{1} \cdot \beta^*S\mathbf{1} < \alpha^*T\mathbf{1} \cdot \beta^*(\zeta\beta)\mathbf{1} \tag{13.11}$$
$$\Longleftrightarrow \quad \alpha^*\eta \cdot \beta^*(1 - \zeta) < \alpha^*(1 - \eta) \cdot \beta^*\zeta$$
$$\Longleftrightarrow \quad \alpha^*\eta < \beta^*\zeta$$

By corollary 13.6 this is equivalent to

$$\mathbb{E}(B) < \mathbb{E}(A)$$

which is our usual condition that the mean service time is strictly smaller than the mean inter–arrival time.

Notes

Discrete PH distributions have been introduced in Neuts [62, 65]. For an early application and discrete time MAPs see Alfa and Neuts [4]. A text book presentation can be found in Latouche and Ramaswami [52]. Discrete phase–type distributions with infinitely many phases are introduced in Shi and Liu [79]. The analysis of the GI/G/1 queue in discrete time is taken from Alfa and Li [3] and Alfa [2]. An overview on further results is given in Alfa [1].

Exercise 13.1 Prove lemma 13.2.

Exercise 13.2 Verify the remaining and elapsed time representations introduced in example 13.5.

Exercise 13.3 Prove formula (13.3) for the factorial moments of a discrete phase–type distribution.

Exercise 13.4 For matrices A, B, and C of appropriate dimensions, prove the distributive laws

$$A \otimes C + B \otimes C = (A + B) \otimes C$$
$$A \otimes B + A \otimes C = A \otimes (B + C)$$

for the Kronecker product, and verify equalities (13.8) and (13.9).

Exercise 13.5 For matrices A, B, C, and D of appropriate dimensions, prove the associative law

$$(A \otimes B)(C \otimes D) = AC \otimes BD$$

for the Kronecker product, and verify formula (13.10) and the equivalence (13.11).

Exercise 13.6 Analogous to an interrupted Poisson process (IPP), we define an **interrupted Bernoulli process**. There are two phases (numbered 1 for "on" and 0 for "off"). In every time slot there is a probability p to switch from phase 0 to phase 1 and a probability q to switch back. In phase 1 there is a probability r of observing one arrival in a time slot. Give an exact definition in terms of a transition matrix for a discrete MAP. Note the difference to the IPP due to the possibility of phase change and arrival occuring in the same time slot.

Chapter 14

SPATIAL MARKOVIAN ARRIVAL PROCESSES

With respect to queueing theory we have, for several times, pointed to the application area of telecommunication networks. In fact, during the last two decades the analysis of complex systems of that type has become the most significant issue in applied queueing theory. Modern communication facilities represent articles of daily use, and are outfit accessories of pedestrians, car drivers, pilots, and nearly all people who need the contact to other people or to data processing devices. Mobility and spatial distribution are characteristic features of these systems. The installation of mobile communication networks (that started in a technically useful form as early as in 1982[1]) is accompanied in many cases by a partition of a geographic region into cells covering the whole area. Customers of such networks get active randomly in time, and are moving around in and across the cells, eventually stopping their activities (vanishing as network users) after having been serviced by the providing company.

Transferred into the language of queueing theory we are confronted thereby with a new type of arrival process and a new species of customers, namely processes that put their arriving elements (customers) onto certain locations, and customers who start moving immediately after appearing at a location. Arrival processes of that kind are characterized by a random behavior in time *and space*.

In previous chapters we have stepped through various queueing models until reaching types "beyond the exponential", and we saw that Markovian arrival processes (MAPs) belong to the most versatile tools for describing the dynamics of modern computer networks (and related configurations). What we

---

[1] The Advanced Mobile Phone System (AMPS), developed by Bell Laboratories, USA.

are now about to do is a generalization of these processes to spatial arrival processes.

1. Arrivals in Space

The Markovian arrival processes $(N_t, J_t)_{t \geq 0}$ considered so far had state space $\mathbb{N}_0 \times E$, where $E = \{1, \ldots, m\}$ represented the phase space, N_t a counting variable, and $(J_t)_{t \geq 0}$ a time-homogeneous Markov process. Nothing was said about "where" an arrival occurs, or what kind of additional information we can assign to the "customers" or "jobs" that arrive according to a MAP.

The properties of $\mathbb{N}_0$ that we needed for an adequate description of the counting variable N_t may be seen as to be the following:

(i) We can measure (count) the "jobs" that arrivals bring into the system (whatever the latter is),

(ii) we can add sets of "jobs" that arrived, i.e. the number of "jobs" that several (possibly not subsequent) arrival events produce is the sum of all individual arrival sets, and

(iii) $(N_t)_{t \geq 0}$ is an increment process with respect to the portions that arrivals add to the system, i.e., for $\alpha \subset \mathbb{N}_0$ with $A = \sum_{\alpha_i \in \alpha} \alpha_i$, and $K \subset J$, we have

$$\mathbb{P}((N_{s+t}, J_{s+t}) \in A \times K \mid N_s = n, J_s = i) =$$
$$\mathbb{P}((N_{s+t}, J_{s+t}) \in (A - n) \times K \mid N_s = 0, J_s = i).$$

This somewhat artificially looking description attains its meaning next when we are going to generalize the concept of a MAP by adding information to the arriving elements (e.g. jobs). In more general terms, namely, a MAP can be regarded as a two-dimensional Markovian jump process $(N_t, J_t)_{t \geq 0}$ with state space $\mathbb{U} \times E$, where E represents the phase process, and $\mathbb{U}$ has the following properties:

1 There is a σ-algebra $\mathcal{U}$ such that $(\mathbb{U}, \mathcal{U})$ is a measurable space with $\{u\} \in \mathcal{U}$ for any $u \in \mathbb{U}$.

2 $(\mathbb{U}, +)$ forms a semi-group with neutral element o.

3 For $A \subset \mathbb{U}$, $K \subset J$, and $A - u = \{v \in \mathbb{U} : v + u \in A\}$,

$$\mathbb{P}((N_{s+t}, J_{s+t}) \in A \times K \mid N_s = u, J_s = i) =$$
$$\mathbb{P}((N_{s+t}, J_{s+t}) \in (A - u) \times K) \mid N_s = o, J_s = i).$$

In fact, any MAP can be regarded as a so-called Markov-additive jump process, being defined in general terms as follows. Let U denote a set with properties 1 - 3, and $E \subset \mathbb{N}_0$.

Definition 14.1 A two-dimensional process $(Y_t, J_t)_{t \geq 0}$ on a state space $U \times E$ is called a **Markov-additive jump process** if (i) $(Y_t, J_t)_{t \geq 0}$ is a Markov process, and (ii) for $s, t \geq 0$, the conditional distribution of $(Y_{s+t} - Y_s, J_{s+t})$, given (Y_s, J_s), depends only on J_s.

It is easy to see that, for any Markov-additive jump process $(Y_t, J_t)_{t \geq 0}$, the (phase) component $(J_t)_{t \geq 0}$ forms a Markov jump process and $(Y_t)_{t \geq 0}$ has conditionally independent increments. That is, given the states of J_{t_ν} for $0 \leq \nu \leq n$, the random variables

$$Y_{t_1} - Y_0, \ Y_{t_2} - Y_{t_1}, \ \dots, Y_{t_n} - Y_{t_{n-1}}$$

are conditionally independent for known $J_0, J_{t_1}, \dots, J_{t_n}$.

Let us now consider Markovian arrival processes in which an arrival event means the appearance of customers at specific locations, or simply the appearance of points in some space $\mathbb{X}$. We may speak of *localizable arrivals* in this case, and of a **spatial arrival process**. The term "spatial" requires some explanation. Being accustomed to think in terms of the Euclidian space, usually everybody takes for granted properties of the space $\mathbb{X}$ that have particular mathematical significance. Such properties are (among others)

(i) $\mathbb{X}$ is a metric space with metric $d : \mathbb{X} \times \mathbb{X} \to \mathbb{R}_0$.

(ii) Any Cauchy sequence $\{x_n\}$ in $\mathbb{X}$ is convergent.[2]

(iii) $\mathbb{X}$ contains a countable dense subset.

A space with these properties is called a **Polish space**. Since (ii) means completeness, and (iii) separability, a Polish space can be defined more precisely as a complete separable metric space $(\mathbb{X}, d)$. In such a space any compact subset is closed, and any isolated point constitutes a closed subset. This is the type of space we take as a basis, i.e. when speaking of localizable arrivals. Accordingly, when using the notion of a *spatial arrival process* we shall constantly refer to arrivals in a *Polish space*.

---

[2]A sequence satisfying $d(x_n, x_m) \to 0$ as $m, n \to \infty$.

The appearance of (finitely many) points in $\mathbb{X}$ can mathematically be interpreted as the occurrence of some certain (finite) counting measure ν. This is due to the fact that counting measures are the primary ingredients of point fields. Let $\mathcal{B}(\mathbb{X})$ denote the Borel σ-algebra of $(\mathbb{X}, d)$, and μ a locally finite measure on $\mathcal{B}(\mathbb{X})$, i.e. a measure with the property that for each $x \in \mathbb{X}$ there is some open vicinity $U(x)$ such that $\mu(U) < \infty$. Assume that μ has the following regularity property:

$$\mu(A) = \sup\{\mu(K) : K \subset A, \; K \text{ compact}\} \quad \text{for any } A \in \mathcal{B}(\mathbb{X}).$$

Then μ is called a **Radon measure**, and if the range of μ is $\mathbb{N}_0$, it is called a (Radon) **counting measure**. Any counting measure defines what we may call a **point field** in $\mathbb{X}$. We know that arrival events in a Markovian arrival process (MAP) occur randomly in time. In a similar way, when assuming that each arrival specifies a set of points (or, in case, a single point) in space, we should naturally propose that these points are located randomly in the space. A spatial MAP, hence, produces random point fields in a space $\mathbb{X}$ over time.

What is the precise mathematical description of a random point field? Let $\mathbb{V}$ denote the family of all counting measures, and let, for each subset $S \in \mathcal{B}(\mathbb{X})$, $\mathbb{V}_n(S)$ be the set of all measures μ with $\mu(S) = n$. The σ-algebra $\mathcal{V}$ that is generated by the family

$$\mathcal{M} = \left\{\mathbb{V}_n(S) : n \in \mathbb{N}_0, S \in \mathcal{B}(\mathbb{X})\right\}$$

(of sets of measures) defines $\mathbb{V}$ as a measurable space $(\mathbb{V}, \mathcal{V})$ of counting measures over the Polish space $(\mathbb{X}, d)$. The σ-algebra $\mathcal{V}$ is rich enough to allow the distinction of single measures, i.e. every singleton $\{\mu\}$ in $\mathbb{V}$ is measurable.[3] The answer to our above question is easy now: Given some probability space $(\Omega, \mathcal{A}, \mathbb{P})$, a **random point field** is nothing else than a measurable mapping $\mathbb{F} : \Omega \to \mathbb{V}$.

Obviously, the family $\mathbb{V}$ of counting measures over $(\mathbb{X}, d)$ forms a semi-group with respect to addition, where the sum $\mu + \nu$ of measures is defined as the measure $(\mu + \nu)(S) = \mu(S) + \nu(S)$ for all $S \in \mathcal{B}(\mathbb{X})$. The neutral element o is the measure that assigns zero points to any subset $S \in \mathcal{B}(\mathbb{X})$. We introduce the following notation: Let $A, B \in \mathcal{V}$; then

$$
\begin{aligned}
A - \nu &= \{\mu \in \mathbb{V} : \mu + \nu \in A\}, \\
\mu + B &= \{\mu + \nu \in \mathbb{V} : \nu \in B\}, \\
A - B &= \{\mu \in \mathbb{V} : \mu + B \subset A\}.
\end{aligned}
$$

---

[3] This is due to the fact that a locally finite measure μ on $(\mathbb{X}, d)$ is determined already by its values $\mu(\{x\})$ on the singletons x in $\mathbb{X}$.

From these properties it is easy to deduce that $\mathbb{V}$ can well be used as the state space of the counting variable of some Markov-additive jump process. As that it is an arrival process for measures. More precisely, we introduce what we call an SMAP for short.

Definition 14.2 A homogeneous Markov-additive jump process $(Y_t, J_t)_{t \geq 0}$ with state space $\mathbb{V} \times E$, where $(\mathbb{V}, \mathcal{V})$ is a measurable space of counting measures over a Polish space $(\mathbb{X}, d)$, is called a **spatial Markovian arrival process** or **SMAP**.

Each jump in an SMAP $(Y_t, J_t)_{t \geq 0}$ is to be interpreted as the arrival of some point field in $\mathbb{X}$ (corresponding to a finite measure ν over $\mathbb{X}$) together with some certain phase transition $i \to j$ $(i, j \in E)$. Y_t is the random variable describing the very point field that is created by superposition of all those locally finite counting measures that arrived under the SMAP up to time t.

For $A \subset \mathcal{V}$ and $K \subset J$, the probabilities

$$\mathbb{P}(Y_t \in A, J_t \in K \mid Y_0 = o, J_0 = i) =: p_t(0, i; A \times K)$$

define what is usually called the **transition kernel** of the process. Accordingly, we call

$$\frac{d}{dt} \mathbb{P}(Y_t \in A, J_t \in K \mid Y_0 = o, J_0 = i) =: q(0, i; A \times K)$$

the transition rate kernel of the SMAP (here we propose $A \times K \neq \{(o, i)\}$). Using this notation we define, for each subset $S \in \mathcal{B}(\mathbb{X})$, the **subset specific transition kernels**

$$P_t(n; S) = (P_t(n, ij; S))_{i,j \in E} = (p_t(0, i; \mathbb{V}_n(S) \times \{j\}))_{i,j \in E} \qquad (14.1)$$

as well as the **subset specific transition rate kernels**

$$D_n(S) = (D_{n;ij}(S))_{i,j \in E} = (q(0, i; \mathbb{V}_n(S) \times \{j\}))_{i,j \in E}. \qquad (14.2)$$

Now we are in the position to define a correlated family of subset specific spatial BMAPs by counting, in any fixed subset $S \in \mathcal{B}(\mathbb{X})$, the points that occur within S according to the arrivals under the SMAP $(Y_t, J_t)_{t \geq 0}$. Since a batch Markovian arrival process, as a Markov process, is completely determined by its generator, it suffices to specify a generator $G(S)$ for any measurable subset S of $\mathbb{X}$. This is done by setting

$$G(S) = \begin{pmatrix} D_0(S) & D_1(S) & D_2(S) & \cdots \\ O & D_0(S) & D_1(S) & \cdots \\ O & O & D_0(S) & \cdots \\ \vdots & \vdots & \cdots & \ddots \end{pmatrix}. \qquad (14.3)$$

The assertion that $G(S)$ is a generator matrix with

$$\sum_{n=0}^{\infty} \sum_{j \in E} D_{n;ij}(S) = 0 \quad \forall \, i \in E$$

is justified by the following lemma.

Lemma 14.3 *For every* $S \in \mathcal{B}(\mathbb{X})$, *the sum* $\sum_{n=0}^{\infty} D_n(S) = D$ *forms the generator matrix of the phase process* $(J_t)_{t \geq 0}$ *of the SMAP* $(Y_t, J_t)_{t \geq 0}$, *independently of S.*

Proof: Excluding transitions of the form $(o, i) \rightarrow (o, i)$ we have, for $j \neq i$,

$$
\begin{aligned}
\sum_{n=0}^{\infty} D_{n;ij}(S) &= \sum_{n=0}^{\infty} q((o, i), \mathbb{V}_n(S) \times \{j\}) \\
&= q\left((o, i), \bigcup_{n \in \mathbb{N}_0} \mathbb{V}_n(S) \times \{j\}\right) \\
&= \frac{d}{dt} p\left(t; (o, i), \bigcup_{n \in \mathbb{N}_0} \mathbb{V}_n(S) \times \{j\}\right) \\
&= \frac{d}{dt} \mathbb{P}\left(J_t = j \mid J_0 = i\right) = D_{ij} \, .
\end{aligned}
$$

On the other hand, for $j = i$,

$$
\begin{aligned}
\sum_{n=0}^{\infty} D_{n;ii}(S) &= D_{ii}(S) = -\sum_{j \neq i} D_{ij}(S) \\
&= -\sum_{j \neq i} \sum_{n=0}^{\infty} D_{n;ij}(S) = -\sum_{j \neq i} D_{ij} = D_{ii} \, ,
\end{aligned}
$$

and hence, $\sum_{n=0}^{\infty} D_n(S) = D$, independently from the choice of $S \in \mathcal{X}$.
$\square$

The BMAP $(N_t(S), J_t)_{t \geq 0}$ with generator matrix $G(S)$ is called a **spatial BMAP over the subset** $S \in \mathcal{B}(\mathbb{X})$, or **SBMAP over** S for short.

Notice, that a common BMAP as introduced in chapter 10 can be seen as an SMAP over a single point $\mathbb{X} = \{x\}$ with time-homogeneous phase process. If $E = \{1\}$, then $(Y_t)_{t \geq 0}$ is a *space-time Poisson process* with general spatial distribution. Such Poisson processes have been considered by Serfozo [77] and Breuer [21].

2. Properties of Spatial MAPs

We call an SMAP *regular*, if $\sum_{(n,j)\neq(0,i)} D_{n;ij}(S) = -D_{0;ii}(S) > 0$, and **stable** if $-D_{0;ii}(S) < \infty$ for any $S \in \mathcal{B}(\mathbb{X})$, i.e. an SMAP is stable if the total arrival rate connected with any phase transition is finite. We assume regularity and stability throughout. Let us ask for the phase depending probabilities $P_t(k; S)$ that k customers have arrived until time epoch t in some subset $S \in \mathcal{B}(\mathbb{X})$. Since, for fixed S, we can proceed as in case of a common BMAP, we immediately obtain

$$P_t(k; S) = \left(e^{*\Delta(S) \cdot t}\right)_k = \sum_{n=0}^{\infty} \frac{t^n}{n!} \Delta(S)_k^{*n},$$

where

$$\Delta(S)_k^{*n} = G(S)_{0\,k}^n \quad \text{for } k \geq 0.$$

As a consequence, the counting variables $N_t(S)$ satisfy

$$\mathbb{P}(N_t(S) = n, J_t = j \mid N_0(S) = 0, J_0 = i) = P_t(n, ij; S). \tag{14.4}$$

We also have $P_t(0; S) = e^{D_0(S) \cdot t}$ as before, and $P_t(n; S) = \left(e^{*\Delta(S) \cdot t}\right)_n$. Written in block matrix form, $P_t(S)$ reads

$$P_t(S) = \begin{pmatrix} e^{D_0(S) \cdot t} & \left(e^{*\Delta(S) \cdot t}\right)_1 & \left(e^{*\Delta(S) \cdot t}\right)_2 & \cdots \\ O & e^{D_0(S) \cdot t} & \left(e^{*\Delta(S) \cdot t}\right)_1 & \cdots \\ O & O & e^{D_0(S) \cdot t} & \cdots \\ \vdots & \vdots & \vdots & \ddots \end{pmatrix}.$$

Justified by our result $\sum_{n=0}^{\infty} D_n(S) = D$, the phase process $(J_t : t \geq 0)$ is the same for any SBMAP over $S \in \mathcal{B}(\mathbb{X})$ and plays the same role as in case of a common BMAP. Hence, the form of the transition matrix P_t^{Φ} of $(J_t : t \geq 0)$ remains unchanged as $P_t^{\Phi} = e^{D \cdot t}$.

Facing this and the above statements it is obvious that nearly all properties of a common BMAP reappear as those of an SMAP when considering only one fixed subset $S \in \mathcal{B}(\mathbb{X})$.

There are two points, but, that have to be emphasized when dealing with an SMAP:

1 We have to determine also joint distributions of points in different subsets $S_\nu, \nu = 1, \ldots, K$, for any given family $\{S_1, \ldots, S_K\} \subset \mathcal{B}(\mathbb{X})$ in order to fully define a spatial arrival process (and to allow realistic applications of the theory).

2 There is a need for recipies for defining the random point fields that occur
   according to arrivals under the SMAP.

Addressing the first point, consider a family of measurable subsets $S_1, \ldots, S_K$
in the Polish space $\mathbb{X}$. The joint distribution of points in such a family that
accumulated due to arrivals under the SMAP can be determined as follows.
Let the sets $S_1, \ldots, S_K \in \mathcal{B}(\mathbb{X})$ be disjoint[4], and set $\mathbf{S} = (S_1, \ldots, S_K)$,
$\mathbf{n} = (n_1, \ldots, n_K)$. We use the notation

$$
\begin{aligned}
\mathbb{V}_{\mathbf{n}}(\mathbf{S}) &= \{\nu \in \mathbb{V} : \nu(S_k) = n_k,\ 1 \leq k \leq K\}, \\
P_t(\mathbf{n}, i, j; \mathbf{S}) &= p(t; i, \mathbb{V}_{\mathbf{n}}(\mathbf{S}) \times \{j\}), \\
D(\mathbf{n}, i, j; \mathbf{S}) &= q(i, \mathbb{V}_{\mathbf{n}}(\mathbf{S}) \times \{j\}), \\
P_t(\mathbf{n}; \mathbf{S}) &= (P_t(\mathbf{n}, i, j; \mathbf{S}))_{i,j \in E}, \\
D(\mathbf{n}; \mathbf{S}) &= (D(\mathbf{n}, i, j; \mathbf{S}))_{i,j \in E}, \\
P_t(\mathbf{S}) &= \{P_t(\mathbf{n}; \mathbf{S})\}_{\mathbf{n} \in \mathbb{N}_0^K}, \\
\Delta(\mathbf{S}) &= \{D(\mathbf{n}; \mathbf{S})\}_{\mathbf{n} \in \mathbb{N}_0^K}.
\end{aligned}
$$

Then, due to conditional independence of increments, the Chapman-Komogorov
equations hold exactly as in the case of a common BMAP, i.e. written in convolutional form,

$$P_{t+\tau}(\mathbf{S}) = P_t(\mathbf{S}) * P_\tau(\mathbf{S}).$$

By subtracting $P_t(\mathbf{S})$ on both sides and forming the differential quotient, we
obtain the Chapman-Kolmogorov differential equations:

$$\frac{d}{dt} P_t(\mathbf{S}) = \Delta(\mathbf{S}) * P_t(\mathbf{S}).$$

Similar to the case of one-dimensional Markov processes as seen in chapter
10, the solution of these equations takes the (convolutional) exponential form:

$$P_t(\mathbf{S}) = e^{*\Delta(\mathbf{S})\, t}, \quad P_t(\mathbf{n}; \mathbf{S}) = \sum_{k=0}^{\infty} \frac{t^k}{k!} (\Delta^{*k}(\mathbf{S}))_{\mathbf{n}}.$$

Thus, the expressions for joint distributions of customer populations in disjoint
subsets formally resemble those of a one-dimensional BMAP. The correspond-
ing expressions for the SMAP with respect to one single subset $S \in \mathcal{B}(\mathbb{X})$ are
obtained from the results for $K = 1$.

Addressing now the second point, one possible method to specify the types of
the random point fields occurring as arrivals under an SMAP is the following.

---

[4]It should be obvious that there is no loss of generality by that assumption, since intersections may be
treated as separate subsets.

Start with some common BMAP that is given in form of its phase process $(J_t : t \geq 0)$ and its rate matrices $D_n = (D_{n;ij})_{i,j \in E}$. Then define a family

$$\Phi = \{\phi_{ij} : i, j \in E\}$$

of probability measures over the Polish space $(\mathbb{X}, d)$, such that, for any $S \in \mathcal{B}(\mathbb{X})$, $\phi_{ij}(S)$ represents the probability that an arriving batch in coincidence with a phase transition from i to j is located in S. To be more concrete, let $p_i(n, j)$ denote the probability that the BMAP, upon changing its phase from i to j, creates a batch of size $n \geq 0$. Then we define

$$p_i(n, j; S) = p_i(n, j)\, \phi_{ij}(S) \quad \text{for all} \quad i, j \in E, \ n \geq 1,$$

$$p_i(0, j; S) = p_i(0, j) + \sum_{n=1}^{\infty} p_i(n, j; \mathbb{X} \setminus S) \quad \text{for all} \ j \neq i.$$

In this notation $p_i(n, j; S)$ is the probability that a batch of size n is located in S in coincidence with a phase transfer from i to j. Let, as usual, N_t be the random number of jobs arrived until t according to the BMAP, and γ_i be the total instantaneous transition rate when the phase is i, i.e.

$$\gamma_i = \sum_{n=0}^{\infty} \sum_{\substack{j \in E \\ j \neq i}} D_{n;ij} + \sum_{n=1}^{\infty} D_{n;ii}, \quad i \in E.$$

The BMAP $(N_t, J_t)_{t \geq 0}$ is completely described by its rate matrices, and so would be our spatial MAP if its rate matrices were given in turn. The latter, now, can be easily realized for each $S \in \mathcal{B}(\mathbb{X})$ by setting

$$D_{0;ii}(S) = -\gamma_i \left(1 - \sum_{n=1}^{\infty} p_i(n, i; \mathbb{X} \setminus S) \right),$$

$$D_{0;ij}(S) = \gamma_i p_i(0, j; S) \quad \text{for } j \neq i,$$

$$D_{n;ij}(S) = \gamma_i p_i(n, j; S) \quad \text{for } n \geq 1.$$

These matrices define the generator of the SBMAP over S and thereby the process itself. The family Φ of probability measures ϕ_{ij} over $\mathbb{X}$ determines *where* to locate a batch, and the size of any batch arriving under the BMAP specifies the number of points at that very location. This is for sure a somewhat restrictive specification of the random point fields (each by intuition may be seen as to be a superposition of points at one location), but since the locations themselves vary according to the measures ϕ_{ij} the method works in practice.

Another method may consist in assigning, to each pair $(i, j) \in E \times E$ some positive integrable fuction $\xi_{ij} : \mathbb{X} \to \mathbb{R}^+$, such that each $S \in \mathcal{B}(\mathbb{X})$ contains a random number $N(S)$ of points with probability

$$\frac{\varphi_{ij}(S)^n}{n!}e^{-\varphi_{ij}(S)},$$

where

$$\varphi_{ij}(S) = \int_S \xi_{ij}(x)dx.$$

This leads us to some simple type of a random Poisson field with mean $\varphi_{ij}(S)$ for any $S \in \mathcal{B}(\mathbb{X})$. The connection to SMAP arrivals is given by setting

$$\begin{aligned}
p_i(n,j;S) &= \frac{\varphi_{ij}(S)^n}{n!}e^{-\varphi_{ij}(S)} \quad \text{for } n \geq 1, \\
p_i(0,j;S) &= e^{-\varphi_{ij}(S)} \quad \text{for } j \neq i, \\
p_i(0,i;S) &= 0.
\end{aligned}$$

Let the total transition rate γ_i out of some state (v,i) for an SMAP be given. Then, assuming $E = \{1, \ldots, m\}$,

$$\begin{aligned}
D_{n;ij}(S) &= \gamma_i\, p_i(n,j;S) \quad \text{for } (n,j) \neq (0,i) \\
D_{0;ii}(S) &= -\gamma_i(m - e^{-\varphi_{ii}(S)}),
\end{aligned}$$

where γ_i is some finite positive constant for every $i \in E$.

Comments on the modelling of customer motion.
It is easy to see that we can describe the movement of customers by time dependent mappings of the Polish space $(\mathbb{X}, d)$ into itself. A customer in a mobile communication system, for example, who is located at time t_0 at a point $\mathbf{x}(t_0)$ when requesting a call, may move from $\mathbf{x}(t_0)$ to $\mathbf{y} = \mathbf{x}(t_1)$ during a time interval $(t_0, t_1]$. That way, if he is active for, say, T time units, he follows some curve $\{\mathbf{x}(s) : s \in [t_0, t_0 + T]\}$ through the landscape and then vanishes from the system — from the point of view of the telecommunication provider — due to call completion. In our models the "landscape" is part of the Polish space $(\mathbb{X}, d)$, and the curve relates the points $\mathbf{x}(t)$ to the user's starting point $\mathbf{x}(t_0)$ according to some rule $\Upsilon_t : \mathbf{x}(t_0) \mapsto \mathbf{x}(t_0 + t) = \mathbf{x}(t)$ for $0 \leq t \leq T$. If we know the parameters $\mathbf{x}(t_0)$ (the arrival location), T (the service time duration), and Υ_t (the rule for the displacement after t time units) for each user, we can decide at any time whether or not there is an active user in the system at some arbitrary location $\mathbf{y}$. The system would be a queueing system in space and time. For its analytical description a pecularity has to be taken into account: In reality, the point mappings $\Upsilon_t : \mathbf{x}(t_0) \mapsto \mathbf{x}(t_0 + t)$ are resembling random walks in most cases since human customers normally behave individually and the curves they follow are random in general, and completely different. One

way to cope with this problem is to prescribe probabilities for the displacement of customers with respect to their arrival locations and arrival times. A
much more simple approach is based on the assumption of *deterministic* motion where the $(\Upsilon_t : t \geq 0)$ form a given topological group of mappings. In
fact, this restriction is not that serious as it may seem at a first glance. On
the one side, in many practical situations one is faced with the task to model
the impact of movements that take place along streets or railway lines, such
that there are streams of uniformly moving individuals or cars or trains subject
to the same deterministic law. On the other side, the superposition of several (finitely many) streams of that type may well mirror an average behaviour
of customers in more complex configurations. Such superpositions, although
causing additional analytical complexity, can be handled in principle without
problems. Let us shortly indicate how to determine time dependent probabilities for the spatial distribution of customers (users of some facility, jobs, etc.)
in a space-time queueing system.

We assume that there is a service process defined that reflects the treatment of
the customers up to their disappearance out of the system. Let $R = (\Upsilon_t : \mathbb{X} \to
\mathbb{X}, \ t \geq 0)$ be an abelian topological group with the topology $\mathcal{O} = \{\Upsilon_s : s \in
O, \ O \text{ open in } (\mathbb{X}, d)\}$. Then the following holds.

1 Given any neighbourhood W of $\Upsilon_t \circ \Upsilon_s$, there are neighbourhoods U and
   V such that $U \circ V \subset W$.

2 For any neighbourhood V of Υ_t^{-1} there is some neighbourhood U with
   $U^{-1} = \{\Upsilon_r^{-1} : \Upsilon_r \in U\} \subset V$.

We write Υ_{t+s} for $\Upsilon_t \circ \Upsilon_s$, and assume that any customer starts moving immediately after arriving at location $\mathbf{x}$ according to the law

$$\Upsilon_s(\mathbf{x}) = \mathbf{x}(s), \quad s \in (0, T], \tag{14.5}$$

where T is the time he spends in the system. The set

$$\Upsilon_s[S] = \{\mathbf{y} = \Upsilon_s(\mathbf{x}) : \mathbf{x} \in S\}, \quad s \geq 0$$

is called the **displacement set** of S for any $S \in \mathcal{B}(\mathbb{X})$. Similarly, the set

$$\Upsilon_{-s}[S] = \{\mathbf{x} : \Upsilon_s(\mathbf{x}) \in S\}, \quad s \geq 0$$

is called the **source set** of S with respect to $\Upsilon_s[S]$. Note that $\Upsilon_{-s}[S] =
(\Upsilon_s)^{-1}[S]$ due to proposed group property. Given a spatial Markovian arrival
process as mentioned above, we are able to compute the probability matrices
$P_t(S)$ describing the phase depending numbers of arrivals up to time t for any

subset $S \in \mathcal{B}(\mathbb{X})$. A customer is called (S, t)-**resident** if, after his arrival somewhere in $\mathbb{X}$ at a time $u \leq t$, his service (that started at u) continues to go on beyond t, and his location at time t is in $S \in \mathcal{B}(\mathbb{X})$. The random number of (S, t)-resident customers observed at time $u \leq t$ is denoted by $N_{u,t}(S)$, such that $N_t(S) = N_{t,t}(S)$ represents the number of all those customers who are located in S at time t. Assume that there is a possibility to determine the probabilities

$$Q_{r;ij}(u, t; S) = \mathbb{P}(N_{u,t}(S) = r, J_u = j \mid N_{0,t}(S) = 0, J_0 = i)$$

that define the distribution of the random variables $N_t(S)$ in case that customers do not move.[5] Then, if movements are allowed and happen according to the law (14.5), the corresponding distribution for $N_t(S)$ is obtained by merely replacing S in $Q_{r;ij}(u, t; S)$ by its source set $\Upsilon_{-(t-u)}[S]$ and performing the same computation.

Notes

A first definition of a spatial batch Markovian arrival processes based on the construction of probability mass functions in the Euclidean space traces back to Baum [10]. Subsequent treatments and generalizations are due to Baum and Kalashnikov [11], and Breuer [21]. In [21] a type of spatial process has been investigated (among others) that is classified as a space-time Poisson process with general spatial distribution. This type has been considered also by Serfozo in [77]. An important application area of spatial BMAPs and corresponding queueing models is the performance analysis of todays telecommunication systems. The handling of customer motion during service in such systems has been treated by Baum and Kalashnikov [12, 13], and Baum and Sztrik [14].

Exercise 14.1 Let the subset specific rate matrices of an SMAP be given by

$$
\begin{aligned}
D_{n;ij}(S) &= \gamma_i\, p_i(n, j; S) \quad \text{for } (n, j) \neq (0, i) \\
D_{0;ii}(S) &= -\gamma_i \left(1 - \sum_{n=1}^{\infty} p_i(n, i; \mathbb{X} \setminus S) \right),
\end{aligned}
$$

where $p_i(n, j; S)$ is determined as the probability for the event that an arrival under a common BMAP occurs in S together with a phase transition $i \to j$, $p_i(n, j; S)$ being defined with means of a family

$$\Phi = \{\phi_{ij} : i, j \in E\}$$

---

[5]Techniques for the computation of the $Q_{r;ij}(u, t; S)$ are presented, for example, in [13].

of probability measures over the Polish space $(\mathbb{X}, d)$ as mentioned in the text above.

The particular arrival rates $\lambda_i(S)$ into S must be dependent upon the choice of subset $S \in \mathcal{B}(\mathbb{X})$. The spatial BMAP that is generated with respect to the fixed chosen subset S can be expressed, for $(n, j) \neq (0, i)$, by its rates $\lambda_i(S)$ and routing probabilities $\pi_i(n, j; S)$ according to $D_{n;ij}(S) = \lambda_i(S) \cdot \pi_i(n, j; S)$, where $\pi_i(n, j; S)$ is the probability that n customers arrive in S together with a phase transition from i to j. Show, that

$$\lambda_i(S) = \gamma_i \left(1 - \sum_{n=1}^{\infty} p_i(n, i; \mathbb{X} \setminus S) \right),$$

$$\pi_i(n, j; S) = \frac{p_i(n, j; S)}{1 - \sum_{n=1}^{\infty} p_i(n, i; \mathbb{X} \setminus S)} \quad \text{for } (n, j) \neq (0, i).$$

Exercise 14.2 Consider the second version of an SMAP realization given in the text above, where each point field that occurs together with a phase transition $i \to j$ is a Poisson point field with mean φ_{ij}. Show that, for any $S \in \mathcal{B}(\mathbb{X})$, $\varphi_{ij}(S) = \mathbb{E}[N(S)]$, where $N(S)$ is the random variable describing the number of points in S.

Chapter 15

APPENDIX

1. Conditional Expectations and Probabilities

Let $(\Omega, \mathcal{A}, P)$ denote a probability space and $(S, \mathcal{B})$ a measurable space. A **random variable** is a measurable mapping $X : \Omega \rightarrow S$, which means that $X^{-1}(B) \in \mathcal{A}$ for all $B \in \mathcal{B}$. In other words, X is a random variable if and only if $X^{-1}(\mathcal{B}) \subset \mathcal{A}$. In stochastic models, a random variable usually gives information on a certain phenomenon, e.g. the number of users in a queue at some specific time.

Consider any real–valued random variable $X : (\Omega, \mathcal{A}) \rightarrow (\mathbb{R}, \mathcal{B})$, $\mathcal{B}$ denoting the Borel σ–algebra on $\mathbb{R}$, which is integrable or non–negative. While the random variable X itself yields the full information, a rather small piece of information on X is given by its **expectation**

$$\mathbb{E}(X) := \int_\Omega X \, dP$$

The conditional expectation is a concept that yields a degree of information which lies between the full information X and its expectation $\mathbb{E}(X)$.

To motivate the definition, we first observe that the distribution $P^X = P \circ X^{-1}$ of X is a measure on the sub–σ–algebra $X^{-1}(\mathcal{B})$ of $\mathcal{A}$, i.e. in order to compute

$$P(X \in B) = P^X(B) = \int_{X^{-1}(B)} dP$$

we need to evaluate the measure P on sets

$$A := X^{-1}(B) \in X^{-1}(\mathcal{B}) \subset \mathcal{A}$$

On the other hand, the expectation $\mathbb{E}(X)$ is an evaluation of P on the set $\Omega = X^{-1}(S)$ only. Thus we can say that the expectation employs P only on the trivial σ–algebra $\{\emptyset, \Omega\}$, while X itself employs P on the σ–algebra $X^{-1}(\mathcal{B})$ generated by X.

Now we take any sub–σ–algebra $\mathcal{C} \subset \mathcal{A}$. According to the Radon–Nikodym theorem there is a random variable $X_0 : \Omega \to S$ with $X^{-1}(\mathcal{B}) = \mathcal{C}$ and

$$\int_C X_0 dP = \int_C X dP \qquad (15.1)$$

for all $C \in \mathcal{C}$. This we call the **conditional expectation** of X under $\mathcal{C}$ and write

$$\mathbb{E}(X|\mathcal{C}) := X_0$$

A conditional expectation is P–almost certainly uniquely determined by (15.1). Typical special cases and examples are

Example 15.1 For $\mathcal{C} = \{\emptyset, \Omega\}$, the conditional expectation equals the expectation, i.e. $\mathbb{E}(X|\mathcal{C}) = \mathbb{E}(X)$. For any σ–algebra $\mathcal{C}$ with $X^{-1}(\mathcal{B}) \subset \mathcal{C}$ we obtain $\mathbb{E}(X|\mathcal{C}) = X$.

Example 15.2 Let I denote any index set and $(Y_i : i \in I)$ a family of random variables. For the σ–algebra $\mathcal{C} = \sigma(\bigcup_{i \in I} Y_i^{-1}(\mathcal{B}))$ generated by $(Y_i : i \in I)$, we write

$$\mathbb{E}(X|Y_i : i \in I) := \mathbb{E}(X|\mathcal{C})$$

By definition we obtain for a σ–algebra $\mathcal{C} \subset \mathcal{A}$, random variables X and Y, and real numbers α and β

$$\mathbb{E}(\alpha X + \beta Y|\mathcal{C}) = \alpha\mathbb{E}(X|\mathcal{C}) + \beta\mathbb{E}(Y|\mathcal{C})$$

For σ–algebras $\mathcal{C}_1 \subset \mathcal{C}_2 \subset \mathcal{A}$ we obtain

$$\mathbb{E}(\mathbb{E}(X|\mathcal{C}_2)|\mathcal{C}_1) = \mathbb{E}(\mathbb{E}(X|\mathcal{C}_1)|\mathcal{C}_2) = \mathbb{E}(X|\mathcal{C}_1) \qquad (15.2)$$

Let $\mathcal{C}_1$ and $\mathcal{C}_2$ denote sub–σ–algebras of $\mathcal{A}$, $\mathcal{C} := \sigma(\mathcal{C}_1 \cup \mathcal{C}_2)$, and X an integrable random variable. If $\sigma(X^{-1}(\mathcal{B}) \cup \mathcal{C}_1)$ and $\mathcal{C}_2$ are independent, then

$$\mathbb{E}(X|\mathcal{C}) = \mathbb{E}(X|\mathcal{C}_1)$$

If X and Y are integrable random variables and $X^{-1}(\mathcal{B}) \subset \mathcal{C}$, then

$$\mathbb{E}(XY|\mathcal{C}) = X \cdot \mathbb{E}(Y|\mathcal{C}) \qquad (15.3)$$

Conditional probabilities are special cases of conditional expectations. Define the **indicator function** of a measurable set $A \in \mathcal{A}$ by

$$1_A(x) := \begin{cases} 1, & x \in A \\ 0, & x \notin A \end{cases}$$

Such a function is a random variable, since

$$1_A^{-1}(\mathcal{B}) = \{\emptyset, A, A^c, \Omega\} \subset \mathcal{A}$$

with $A^c := \Omega \setminus A$ denoting the complement of the set A. Let $\mathcal{C}$ denote a sub–σ–algebra of $\mathcal{A}$. The conditional expectation of 1_A is called **conditional probability** of A. We write

$$P(A|\mathcal{C}) := \mathbb{E}(1_A|\mathcal{C})$$

Immediate properties of conditional probabilities are

$$0 \leq P(A|\mathcal{C}) \leq 1, \qquad P(\emptyset|\mathcal{C}) = 0, \qquad P(\Omega|\mathcal{C}) = 1$$

$$A_1 \subset A_2 \Longrightarrow P(A_1|\mathcal{C}) \leq P(A_2|\mathcal{C})$$

all of which hold P–almost certainly. For a sequence $(A_n : n \in \mathbb{N})$ of disjoint measurable sets, i.e. $A_n \in \mathcal{A}$ for all $n \in \mathbb{N}$ and $A_i \cap A_j = \emptyset$ for $i \neq j$, we obtain

$$P\left(\bigcup_{n=1}^{\infty} A_n \Big| \mathcal{C}\right) = \sum_{n=1}^{\infty} P(A_n|\mathcal{C})$$

P–almost certainly. Let $X : (\Omega, \mathcal{A}) \to (\mathbb{R}, \mathcal{B})$ denote a non–negative or integrable random variable and $Y : (\Omega, \mathcal{A}) \to (\Omega', \mathcal{A}')$ a random variable. Then there is a measurable function $g : (\Omega', \mathcal{A}') \to (\mathbb{R}, \mathcal{B})$ with

$$\mathbb{E}(X|Y) = g \circ Y$$

This is P^Y–almost certainly determined by

$$\int_{A'} g \, dP^Y = \int_{Y^{-1}(A')} X \, dP$$

for all $A' \in \mathcal{A}'$. Then we can define the conditional probability of X given $Y = y$ as $g(y)$. We write

$$\mathbb{E}(X|Y = y) := g(y)$$

for all $y \in \Omega'$.

2.   Extension Theorems

Throughout this book, our basic stochastic tools are either sequences of random variables (such as Markov chains or Markov renewal chains) or even uncountable families of random variables (such as Markov processes, renewal processes, or semi–regenerative processes). It is essential for our models that these random variables are dependent, and in fact we define them in terms of conditional probabilities, i.e. via their dependence structure.

It is then an immediate question whether a probability measure $\mathbb{P}$ exists that satisfies all the postulates in the definition of a stochastic sequence or process. This question is vital as it concerns the very existence of the tools we are using.

2.1   Stochastic chains

Let $(S, \mathcal{B})$ denote a measurable space, μ a probability measure on $(S, \mathcal{B})$, and P_n, $n \in \mathbb{N}$, stochastic **kernels** on $(S, \mathcal{B})$. The latter means that for every $n \in \mathbb{N}$, $P_n : S \times \mathcal{B} \to [0, 1]$ is a function that satisfies
(K1) For every $x \in S$, $P_n(x, .)$ is a probability measure on $(S, \mathcal{B})$.
(K2) For every $A \in \mathcal{B}$, the function $P_n(., A)$ is $\mathcal{B}$–measurable.

Define S^∞ as the set of all sequences $x = (x_n : n \in \mathbb{N}_0)$ with $x_n \in S$ for all $n \in \mathbb{N}_0$. A subset of S^∞ having the form

$$C_{n_1,\ldots,n_k}(A) = \{x \in S^\infty : (x_{n_1}, \ldots, x_{n_k}) \in A\}$$

with $k \in \mathbb{N}$, $n_1 < \ldots < n_k \in \mathbb{N}_0$, and $A \in \mathcal{B}^k$, is called **cylinder** (with coordinates $n_1, \ldots, n_k$ and base A). The set $\mathcal{C}$ of all cylinders in S^∞ forms an algebra of sets. Define $\mathcal{B}^\infty := \sigma(\mathcal{C})$ as the minimal σ–algebra containing $\mathcal{C}$.

Now we can state the extension theorem for sequences of random variables, which is proven in Gikhman and Skorokhod [37], section II.4.

Theorem 15.3 *There is a probability measure $\mathbb{P}$ on $(S^\infty, \mathcal{B}^\infty)$ satisfying*

$$\mathbb{P}(C_{0,\ldots,k}(A_0 \times \ldots \times A_k)) = \int_{A_0} d\mu(x_0) \int_{A_1} P_1(x_0, dx_1) \ldots$$

$$\ldots \int_{A_{k-1}} P_{k-1}(x_{k-2}, dx_{k-1}) \, P_k(x_{k-1}, A_k) \quad (15.4)$$

for all $k \in \mathbb{N}_0$, $A_0, \ldots, A_k \in \mathcal{B}$. The measure $\mathbb{P}$ is uniquely determined by the system (15.4) of equations.

The first part of the theorem above justifies our definitions of Markov chains and Markov renewal chains. The second part states in particular that a Markov chain is uniquely determined by its initial distribution and its transition matrix.

Based on this result, we may define a **stochastic chain** with state space S as a sequence $(X_n : n \in \mathbb{N}_0)$ of S–valued random variables which are distributed according to a probability measure $\mathbb{P}$ on $(S^\infty, \mathcal{B}^\infty)$.

2.2 Stochastic processes

Let S denote a Polish (i.e. a complete separable metric) space, and $\mathcal{B}$ the Borel σ–algebra on S. Define Ω as the set of all functions $f : \mathbb{R}_0^+ \to S$. In order to construct an appropriate σ–algebra on Ω, we again start from the cylinder sets

$$C_{t_1,\ldots,t_k}(A) = \{f \in \Omega : (f(t_1), \ldots, f(t_k)) \in A\}$$

for $k \in \mathbb{N}$, $t_1 < \ldots < t_k \in \mathbb{R}_0^+$, and $A \in \mathcal{B}^k$. Denote the set of all cylinders in Ω by $\mathcal{C}$. Again, $\mathcal{C}$ forms an algebra of sets and we can define $\mathcal{A} := \sigma(\mathcal{C})$ as the minimal σ–algebra containing $\mathcal{C}$.

Let $\mathcal{M} = \{\mu_{t_1,\ldots,t_k} : k \in \mathbb{N}, t_1, \ldots, t_k \in \mathbb{R}_0^+\}$ denote a family of probability distributions with
(C1) For all $k \in \mathbb{N}$, $t_1, \ldots, t_k \in \mathbb{R}_0^+$, and $A \in \mathcal{B}^k$

$$\mu_{t_1,\ldots,t_k,t_{k+1}}(A \times S) = \mu_{t_1,\ldots,t_k}(A)$$

(C2) For all $k \in \mathbb{N}$ and permutations $\pi : \{1, \ldots, k\} \to \{1, \ldots, k\}$

$$\mu_{\pi(t_1,\ldots,t_k)}(\pi(A)) = \mu_{t_1,\ldots,t_k}(A)$$

Then the family $\mathcal{M}$ is called **compatible**.

Remark 15.4 Condition (C1) ensures that the distributions are consistent with each other, while condition (C2) is merely notational.

The following extension theorem by Kolmogorov is proven in Gikhman and Skorokhod [39], section 3.2.

Theorem 15.5 *Let* $\{\mu_{t_1,\ldots,t_k} : k \in \mathbb{N}, t_1, \ldots, t_k \in \mathbb{R}_0^+\}$ *denote a compatible family of probability measures. Then there is a probability measure* $\mathbb{P}$ *on* $(\Omega, \mathcal{A})$ *with*

$$\mathbb{P}(\{f \in \Omega : (f(t_1), \ldots, f(t_k)) \in A\}) = \mu_{t_1,\ldots,t_k}(A) \tag{15.5}$$

for all $k \in \mathbb{N}$, $t_1, \ldots, t_k \in \mathbb{R}_0^+$, *and* $A \in \mathcal{B}^k$. *The measure* $\mathbb{P}$ *is uniquely determined by the system (15.5) of equations.*

Based on this, we define a **stochastic process** with Polish state space S as a family $X = (X_t : t \in \mathbb{R}_0^+)$ of S–valued random variables which are distributed according to a probability measure $\mathbb{P}$ on $(\Omega, \mathcal{A})$. An element $\omega \in \Omega$

is an arbitrary function $\omega : \mathbb{R}_0^+ \to S$. It is also called a **path** of X. If we want to state that the support of $\mathbb{P}$ consists of a special class of functions (say right–continuous ones), then we say that X is a stochastic process with right–continuous paths. The above family $\mathcal{M}$ of probability measures is called the set of **finite–dimensional marginal distributions** for X.

Due to theorem 15.5 a Markov process is uniquely defined by its initial distribution and the family of transition probabilities, since they determine all finite–dimensional marginal distributions. Further our constructions of Markov processes, renewal processes, and semi–Markov processes yield compatible sets of finite–dimensional marginal distributions, hence by theorem 15.5 a probability measure $\mathbb{P}$ for the respective process.

3. Transforms

In several parts of the present book, it is essential to argue via transforms of distributions. The necessary background for these shall be presented shortly in this section. For discrete distributions on $\mathbb{N}_0$ we will introduce z–transforms, while for distributions on $\mathbb{R}_0^+$ the Laplace–Stieltjes transform will be useful.

3.1 z–transforms

Let X denote a $\mathbb{N}_0$–valued random variable with distribution $A = (a_n : n \in \mathbb{N}_0)$, i.e. $\mathbb{P}(X = n) = a_n$ for all $n \in \mathbb{N}_0$. Then the power series

$$A^*(z) := \sum_{n=0}^{\infty} a_n z^n \qquad (15.6)$$

converges absolutely for $z \in \mathbb{C}$ with $|z| \leq 1$ and is analytic in this region. We note that $A^*(z) = \mathbb{E}(z^X)$. If $A(z)$ is a given power series for a distribution $(a_n : n \in \mathbb{N}_0)$, then the probabilities a_n can be derived as

$$a_n = \frac{1}{n!} \frac{d^n}{dz^n} A(z) \Big|_{z=0}$$

for all $n \in \mathbb{N}_0$. Thus the mapping between discrete distributions on $\mathbb{N}_0$ and the power series in (15.6) is bijective, and we may call $A^*(z)$ the (uniquely determined) **z–transform** of X (also: of the distribution A).

Example 15.6 For a Dirac distribution on $k \in \mathbb{N}_0$ with

$$a_n = \begin{cases} 1, & n = k \\ 0, & n \neq k \end{cases}$$

we obtain $A^*(z) = z^k$.

Example 15.7 Let A denote the geometric distribution with some parameter $p \in]0, 1[$, i.e.

$$a_n = (1 - p)p^n$$

for all $n \in \mathbb{N}_0$. The z–transform of A is given by

$$A^*(z) = (1 - p) \sum_{n=0}^{\infty} p^n z^n = \frac{1 - p}{1 - pz}$$

for all $|z| \leq 1$.

A very useful feature is the behaviour of the z–transform with respect to the convolution of two distributions. Let $A = (a_n : n \in \mathbb{N}_0)$ and $B = (b_n : n \in \mathbb{N}_0)$ denote two distributions on $\mathbb{N}_0$. The convolution $C = A * B$ of A and B is defined as the distribution $C = (c_n : n \in \mathbb{N}_0)$ with

$$c_n = \sum_{k=0}^{n} a_k b_{n-k}$$

for all $n \in \mathbb{N}_0$. For the z–transform of C we obtain

$$C^*(z) = \sum_{n=0}^{\infty} c_n z^n = \sum_{n=0}^{\infty} \sum_{k=0}^{n} a_k b_{n-k} z^n = \sum_{n=0}^{\infty} a_k z^k \sum_{n=k}^{\infty} b_{n-k} z^{n-k}$$
$$= A^*(z) \cdot B^*(z)$$

for all $|z| \leq 1$.

This means that the z–transform of a convolution $A * B$ equals the product $A^*(z) \cdot B^*(z)$ of the z–transform of A and B. In terms of random variables we have the following representation: Let X and Y denote two independent $\mathbb{N}_0$–valued random variables. Then the z–transform of the sum $X + Y$ equals the product of the z–transforms of X and Y, i.e.

$$\mathbb{E}\left(z^{X+Y}\right) = \mathbb{E}\left(z^X\right) \cdot \mathbb{E}\left(z^Y\right)$$

for all $|z| \leq 1$.

3.2 Laplace–Stieltjes transforms

Let X denote an $\mathbb{R}_0^+$–valued random variable with distribution function F. The **Laplace–Stieltjes transform** (LST) of X (or: of F) is defined by

$$F^*(s) := \int_0^{\infty} e^{-st} dF(t) = \mathbb{E}\left(e^{-sX}\right)$$

for all $s \in \mathbb{C}$ with $Re(s) \geq 0$. The LST uniquely determines its underlying distribution.

Example 15.8 Let X be exponentially distributed with parameter λ, i.e. X has the distribution function $F(t) = 1 - e^{-\lambda t}$ with Lebesgue density $f(t) = \lambda e^{-\lambda t}$. Then

$$F^*(s) = \int_0^\infty e^{-st} \lambda e^{-\lambda t} \, dt = \frac{\lambda}{s + \lambda}$$

for $Re(s) \geq 0$.

Example 15.9 For the Dirac distribution δ_x on $x \in \mathbb{R}_0^+$ we obtain

$$\delta_x^*(s) = \int_0^\infty e^{-st} dF(t) \qquad \text{with} \qquad F(t) = \begin{cases} 0, & t < x \\ 1, & t \geq x \end{cases}$$

and hence

$$\delta_x^*(s) = e^{-sx}$$

for $Re(s) \geq 0$.

Like the z–transform, the LST is very useful for dealing with convolutions. Let X and Y denote two independent $\mathbb{R}_0^+$–valued random variables. Then the LST of the sum $X + Y$ equals the product of the LSTs of X and Y, i.e.

$$\mathbb{E}\left(e^{-s(X+Y)}\right) = \mathbb{E}\left(e^{-sX}\right) \cdot \mathbb{E}\left(e^{-sY}\right)$$

for all $s \in \mathbb{C}$ with $Re(s) \geq 0$.

Notes

For more on z–transforms see e.g. Juri [43], or the collection of results in Kleinrock [50], appendix I. For Laplace–Stieltjes transforms see chapter XIII in Feller [35] or again Kleinrock [50], appendix I.

4.   Gershgorin's Circle Theorem

An important theorem to find bounds for the eigenvalues of a matrix has been developed by Gershgorin in 1938. For ease of reference it shall be presented in this section. Let $A = (a_{ij})_{i,j \leq m}$ denote a square matrix of dimension m with entries $a_{ij} \in \mathbb{C}$. The following theorem is called **Gershgorin's circle theorem**.

Theorem 15.10 *All eigenvalues of the matrix A lie in the union* $C := \bigcup_{i=1}^{m} C_i$
of the circles

$$C_i = \left\{ z \in \mathbb{C} : |z - a_{ii}| \le \sum_{k \neq i} |a_{ik}| \right\}$$

Proof: Let $x^{(\nu)}$ denote an eigenvector to the eigenvalue λ_ν of A, i.e. $Ax^{(\nu)} = \lambda_\nu x^{(\nu)}$. This implies

$$\sum_{k=1}^{m} a_{ik} x_k^{(\nu)} = \lambda_\nu x_i^{(\nu)} \tag{15.7}$$

for all $i \le m$. Since an eigenvector is determined only up to a scalar multi-plicative, we can assume without loss of generality that there is a component

$$x_{i_0}^{(\nu)} = \max_{1 \le j \le m} \left| x_j^{(\nu)} \right| = 1$$

of the vector $x^{(\nu)}$. Now (15.7) yields for $i = i_0$ the relation

$$\sum_{k \neq i_0} a_{i_0,k} x_k^{(\nu)} = (\lambda_\nu - a_{i_0,i_0}) x_{i_0}^{(\nu)} = \lambda_\nu - a_{i_0,i_0}$$

which implies by the triangle inequality

$$|\lambda_\nu - a_{i_0,i_0}| \le \sum_{k \neq i_0} |a_{i_0,k}| \cdot \left| x_k^{(\nu)} \right| \le \sum_{k \neq i_0} |a_{i_0,k}|$$

Since every eigenvalue satisfies at least one such inequality, the proof is com-plete.
□

Corollary 15.11 *If A is diagonally dominated, i.e. if*

$$|a_{ii}| > \sum_{k \neq i} |a_{ik}|$$

holds for all $1 \le i \le m$, *then the matrix A is invertible.*

Proof: The strict inequality of the assumption implies that $a_{ii} \neq 0$ for all $i \le m$. Applying theorem 15.10 yields a restriction

$$|\lambda| \ge |a_{ii}| - |a_{ii} - \lambda| \ge |a_{ii}| - \sum_{k \neq i} |a_{ik}| > 0$$

for every eigenvalue λ of A. Therefore the matrix A has no eigenvalue zero and thus is invertible.
□

References

[1] A. S. Alfa. Discrete Time Queues and Matrix–analytic Methods. *Top*, 10(2):147–210, 2002.

[2] A. S. Alfa. Combined Elapsed Time and Matrix–Analytic Method for the Discrete Time GI/G/1 and GI^X/G/1 Systems. *Queueing Systems*, 45(1):5–25, 2003.

[3] A. S. Alfa and W. Li. Matrix-geometric analysis of the discrete time $GI/G/1$ system. *Stoch. Models*, 17(4):541–554, 2001.

[4] A. S. Alfa and M. F. Neuts. Modelling vehicular traffic using the discrete time Markovian arrival process. *Transp. Sci.*, 29(2):109–117, 1995.

[5] S. Asmussen. *Applied Probability and Queues*. New York etc.: Springer, 2003.

[6] S. Asmussen and G. Koole. Marked point processes as limits of Markovian arrival streams. *J. Appl. Probab.*, 30(2):365–372, 1993.

[7] S. Asmussen, O. Nerman, and M. Olsson. Fitting phase-type distributions via the EM algorithm. *Scand. J. Stat.*, 23(4):419–441, 1996.

[8] F. Baskett, K. M. Chandy, R. R. Muntz, and F. G. Palacios. Open, closed, and mixed networks of queues with different classes of customers. *Journal of the ACM*, 22:248–260, 1975.

[9] D. Baum. A $BMAP|G|1$-analysis based on convolution calculus. *J. Math. Sci., New York*, 92(4):3990–4002, 1998.

[10] D. Baum. On markovian spatial arrival processes for the performance analysis of mobile communication networks. Technical Report 98–07, University of Trier, Subdept. of Computer Science, 1998.

[11] D. Baum and V. Kalashnikov. Spatial generalization of BMAPs with finite state space. *J. Math. Sci., New York*, 105(6):2504–2514, 2001.

[12] D. Baum and V. V. Kalashnikov. Stochastic models for communication networks with moving customers. *Information Processes*, 1:1–23, 2001.

[13] D. Baum and V. Kalashnikov. Spatial No–Waiting Stations with Moving Customers. *Queueing Systems*, 46:231–247, 2004.

[14] D. Baum and J. Sztrik. Customer motion in queueing models: The use of tangent vector fields. *International Journal on Pure and Applied Mathematics*, 2002. To appear.

[15] G. Bolch, S. Greiner, H. de Meer, and K. S. Trivedi. *Queueing networks and Markov chains*. John Wiley & Sons, New York, Chichester, Weinheim, Brisbane, Singapore, Toronto, 1998.

[16] L. Breiman. *Probability*. Philadelphia, PA: SIAM, 1968.

[17] L. Breuer. The Periodic BMAP/PH/c Queue. *Queueing Systems*, 38(1):67–76, 2001.

[18] L. Breuer. An EM Algorithm for Batch Markovian Arrival Processes and its Comparison to a Simpler Estimation Procedure. *Annals of Operations Research*, 112:123–138, 2002.

[19] L. Breuer. On Markov–Additive Jump Processes. *Queueing Systems*, 40(1):75–91, 2002.

[20] L. Breuer. On the MAP/G/1 Queue with Lebesgue–dominated Service Time Distribution and LCFS Preemptive Repeat Service Discipline. *Stochastic Models*, 18(4):589–595, 2002.

[21] L. Breuer. *From Markov Jump Processes to Spatial Queues*. Kluwer, Dordrecht (Netherlands), 2003.

[22] L. Breuer. Two Examples for Computationally Tractable Periodic Queues. *International Journal of Simulation*, 3(3–4):15–24, 2003.

[23] L. Breuer, A. Dudin, V. Klimenok, and G. Tsarenkov. A Two–Phase BMAP/G/1/N → PH/1/M-1 System with Blocking. *Automation and Remote Control*, 65(1):104–115, 2004.

[24] J. P. Buzen. Computational algorithms for closed queueing networks with exponential servers. *Commun. ACM*, 16:527–531, 1973.

[25] E. Çinlar. *Introduction to stochastic processes*. Englewood Cliffs, N. J.: Prentice-Hall, 1975.

[26] K. M. Chandy, J. H. Howard, and D. F. Towsley. Product form and local balance in queueing networks. *Journal of the ACM*, 24:250–263, 1977.

[27] K. Chung. *Markov chains with stationary transition probabilities*. Springer-Verlag, Berlin etc., 1960.

[28] J. Cohen. *The Single Server Queue*. North–Holland, Amsterdam etc., 1969.

[29] H. Daduna. *Queueing networks with discrete time scale*. Springer, Berlin, Heidelberg, New York, 2001.

[30] N. M. Van Dijk. *Queueing Networks and Product Forms*. John Wiley & Sons, Chichester, New York, Brisbane, Toronto, Singapore, 1993.

[31] J. Doob. *Stochastic processes*. New York: Wiley, 1953.

[32] T. Engset. Emploi du calcul des probabilités pour la détermination du nombre de sélecteurs dans les Bureaux Téléphoniques Centraux. *Rev. gén. elect.*, 9:138–140, 1921.

[33] A. Erlang. Solution of Some Probability Problems of Significance for Automatic Telephone Exchanges. *Electroteknikeren*, 13:5–13, 1917.

[34] W. Feller. *An introduction to probability theory and its applications. Vol. I.* New York etc.: John Wiley and Sons, 1950.

[35] W. Feller. *An introduction to probability theory and its applications. Vol. II. 2nd ed.* New York etc.: John Wiley and Sons, 1971.

[36] E. Gelenbe and G. Pujolle. *Introduction to queueing networks.* John Wiley & Sons, hichester, New York, Brisbane, Toronto, Singapore, 1987.

[37] I. Gihman and A. Skorohod. *The theory of stochastic processes I.* Berlin etc.: Springer, 1974.

[38] I. Gihman and A. Skorohod. *The theory of stochastic processes II.* Berlin etc.: Springer-Verlag, 1975.

[39] I. Gikhman and A. Skorokhod. *Introduction to the theory of random processes.* Saunders, 1969.

[40] W. Gordon and G. Newell. Closed queuing systems with exponential servers. *Oper. Res.*, 15:254–265, 1967.

[41] P. G. Harrison and N. M. Patel. *Performance Modelling of Communication Networks and Computer Architectures.* Addison-Wesley, Reading, Massachusetts, 1993.

[42] J. R. Jackson. Jobshop-like queueing systems. *Management Science*, 10:131–142, 1963.

[43] E. Juri. *Theory and Application of the z–Transform Method.* Wiley, New York, 1964.

[44] V. Kalashnikov. *Topics on Regenerative Processes.* CRC Press, 1994.

[45] K. Kant. *Introduction to computer system performance evaluation.* McGraw-Hill, Inc., New York, London, 1992.

[46] S. Karlin and H. M. Taylor. *A first course in stochastic processes. 2nd ed.* New York etc.: Academic Press, 1975.

[47] S. Karlin and H. M. Taylor. *A second course in stochastic processes.* New York etc.: Academic Press, 1981.

[48] F. Kelly. *Reversibility and Stochastic Networks.* Wiley, 1979.

[49] D. Kendall. Stochastic Processes Occuring in the Theory of Queues and Their Analysis by the Method of the Embedded Markov Chain. *Annals of Mathematical Statistics*, 24:338–354, 1953.

[50] L. Kleinrock. *Queueing systems. Vol. I: Theory.* New York etc.: John Wiley & Sons, 1975.

[51] Y. Lam. A Note on the Optimal Replacement Problem. *Adv. Appl. Prob.*, 20:479–482, 1988.

[52] G. Latouche and V. Ramaswami. *Introduction to matrix analytic methods in stochastic modeling.* Philadelphia, PA: SIAM, 1999.

[53] J. Little. A Proof for the Queueing Formula $L = \lambda W$. *Operations Research,* 9, 1961.

[54] D. M. Lucantoni. New results on the single server queue with a batch Markovian arrival process. *Commun. Stat., Stochastic Models,* 7(1):1–46, 1991.

[55] D. M. Lucantoni. The BMAP/G/1 Queue: A Tutorial. In L. Donatiello and R. Nelson, editor, *Models and Techniques for Performance Evaluation of Computer and Communication Systems,* pages 330–358. Springer, 1993.

[56] D. M. Lucantoni, K. S. Meier-Hellstern, and M. F. Neuts. A single-server queue with server vacations and a class of non-renewal arrival processes. *Adv. Appl. Probab.,* 22(3):676–705, 1990.

[57] D. M. Lucantoni and M. F. Neuts. Simpler proofs of some properties of the fundamental period of the MAP/G/1 queue. *J. Appl. Probab.,* 31(1):235–243, 1994.

[58] F. Machihara. A New Approach to the Fundamental Period of a Queue with Phase–type Markov Renewal Arrivals. *Stochastic Models,* 6(3):551–560, 1990.

[59] S. Meyn and R. Tweedie. *Markov chains and stochastic stability.* Berlin: Springer-Verlag, 1993.

[60] I. Mitrani. *Modelling of computer and communication systems.* Cambridge University Press, Cambridge etc., 1987.

[61] R. Nelson. *Probability, stochastic processes, and queueing theory.* New York, NY: Springer-Verlag, 1995.

[62] M. F. Neuts. Probability distributions of phase type. In *Liber Amicorum Prof. Emeritus H. Florin,* pages 173–206. Department of Mathematics, University of Louvain, Belgium, 1975.

[63] M. F. Neuts. Markov chains with applications in queueing theory, which have a matrix-geometric invariant probability vector. *Adv. Appl. Probab.,* 10:185–212, 1978.

[64] M. F. Neuts. A versatile Markovian point process. *J. Appl. Probab.,* 16:764–774, 1979.

[65] M. F. Neuts. *Matrix–Geometric Solutions in Stochastic Models.* Johns Hopkins University Press, Baltimore, 1981.

[66] M. F. Neuts. *Structured stochastic matrices of M/G/1 type and their applications.* New York etc.: Marcel Dekker, 1989.

[67] M. F. Neuts and J.-M. Li. An algorithm for the $P(n, t)$ matrices of a continuous BMAP. In *Matrix-analytic methods in stochastic models (Flint, MI),* volume 183 of *Lecture Notes in Pure and Appl. Math.,* pages 7–19. Dekker, New York, 1997.

[68] A. Pacheco and N. Prabhu. Markov-additive processes of arrivals. In J. H. Dshalalow, editor, *Advances in queueing,* pages 167–194. CRC Press, Boca Raton, FL, 1995.

[69] R. Pyke. Markov Renewal Processes: Definitions and Preliminary Properties. *Annals of Mathematical Statistics,* 32:1231–1242, 1961.

[70] R. Pyke. Markov Renewal Processes with Finitely Many States. *Annals of Mathematical Statistics*, 32:1243–1259, 1961.

[71] V. Ramaswami. The $N/G/1$ queue and its detailed analysis. *Adv. in Appl. Probab.*, 12(1):222–261, 1980.

[72] V. Ramaswami. A stable recursion for the steady state vector in Markov chains of M/G/1 type. *Commun. Stat., Stochastic Models*, 4(1):183–188, 1988.

[73] M. Reiser and S. Lavenberg. Mean-value analysis of closed multichain queuing networks. *J. Assoc. Comput. Mach.*, 27:313–332, 1980.

[74] S. Ross. *Applied Probability Models with Optimization Applications*. Holden–Day, San Francisco, 1970.

[75] S. Ross. *Stochastic Processes*. John Wiley & Sons, New York etc., 1983.

[76] T. Ryden. An EM algorithm for estimation in Markov-modulated Poisson processes. *Comput. Stat. Data Anal.*, 21(4):431–447, 1996.

[77] R. Serfozo. *Introduction to stochastic networks*. Springer-Verlag, New York, 1999.

[78] D. Shi, J. Guo, and L. Liu. SPH–Distributions and the Rectangle–Iterative Algorithm. In S. R. Chakravarthy and A. S. Alfa, editors, *Matrix–Analytic Methods in Stochastic Models*, pages 207–224, New York etc., 1997. Marcel Dekker.

[79] D. Shi and L. Liu. Markovian Models for Non–negative Random Variables. In A. S. Alfa and S. R. Chakravarthy, editors, *Advances in Matrix Analytic Methods for Stochastic Models*, pages 403–427. Notable Publications, 1998.

[80] W. Smith. Regenerative Stochastic Processes. *Proceedings Royal Society, Series A*, 232:6–31, 1955.

[81] W. Smith. Renewal Theory and Its Ramifications. *Journal of the Royal Statistical Society, Series B*, 20:243–302, 1958.

[82] T. Takine. A new recursion for the queue length distribution in the stationary BMAP/G/1 queue. *Stochastic Models*, 16(2):335–341, 2000.

[83] R. Tweedie. Operator-geometric stationary distributions for Markov chains, with application to queueing models. *Adv. Appl. Probab.*, 14:368–391, 1982.

Index